PLANT SCIENCE RESEARCH AND PRACTICES

PRUNUS

CLASSIFICATION, CULTIVATION AND TOXICITY

Plant Science Research and Practices

Additional books and e-books in this series can be found on Nova's website under the Series tab.

PLANT SCIENCE RESEARCH AND PRACTICES

PRUNUS

CLASSIFICATION, CULTIVATION AND TOXICITY

WILLIAM SCHNEIDER
EDITOR

NOTICE TO THE READER

The Publisher has taken reasonable care in the preparation of this book, but makes no expressed or implied warranty of any kind and assumes no responsibility for any errors or omissions. No liability is assumed for incidental or consequential damages in connection with or arising out of information contained in this book. The Publisher shall not be liable for any special, consequential, or exemplary damages resulting, in whole or in part, from the readers' use of, or reliance upon, this material. Any parts of this book based on government reports are so indicated and copyright is claimed for those parts to the extent applicable to compilations of such works.

Independent verification should be sought for any data, advice or recommendations contained in this book. In addition, no responsibility is assumed by the Publisher for any injury and/or damage to persons or property arising from any methods, products, instructions, ideas or otherwise contained in this publication.

This publication is designed to provide accurate and authoritative information with regard to the subject matter covered herein. It is sold with the clear understanding that the Publisher is not engaged in rendering legal or any other professional services. If legal or any other expert assistance is required, the services of a competent person should be sought. FROM A DECLARATION OF PARTICIPANTS JOINTLY ADOPTED BY A COMMITTEE OF THE AMERICAN BAR ASSOCIATION AND A COMMITTEE OF PUBLISHERS.

Additional color graphics may be available in the e-book version of this book.

Library of Congress Cataloging-in-Publication Data

ISBN: 978-1-53617-755-8

Published by Nova Science Publishers, Inc. † New York

CONTENTS

PREFACE

Prunus: Classification, Cultivation and Toxicity opens with a study wherein functional components of prunes such as caffeoylquinic acid isomers and quinic acid are quantified by means of high-performance liquid chromatographyanalysis.

Prunus avium L is examined, particularly focusing on their profile of bioactive compounds, the health benefits associated with their consumption, as well as their potential therapeutic properties.

Also examined in this collection is *Prunus serotina* Ehrh., a tree species native to North America, appreciated for its valuable wood and ornamental leaves and flowers.

The classification, toxicity, and techniques used to propagate and disseminate the endangered *Prunus africana* are assessed in an effort to improve its medicinal utilization and sustainable conservation.

Chapter 1 - Prunes are the dried fruits of some cultivars of *Prunus domestica* L and contain large amounts of phytochemicals. In this study, functional components in prunes such as caffeoylquinic acid (CQA) isomers and quinic acid were quantified by means of HPLC analysis. It has become apparent that prunes contain relatively high amounts of 4-CQA than other fruits, and the content of quinic acid was about 10 times higher than the values in earlier studies. The contribution of CQA isomers to antioxidant activity of prunes was estimated to be 28.4% based on oxygen radical absorbance capacity (ORAC), hence, it was indicated that residual ORAC is dependent on unknown antioxidant components. Total 39 compounds were isolated from prunes and their chemical structures were elucidated on the basis of NMR and MS analyses. Four sesquiterpenoids, three monoterpene derivatives, a chromanone, and a bipyrrole were novel compounds, and two conformational isomers of 3-CQA were also characterized. Antioxidant activities of isolated compounds were evaluated, and each CQA isomer showed high and similar antioxidant activity on Oil Stability Index method, O_2^- scavenging activity, and ORAC. On the other hand, the activities on ORAC and ferric thiocyanate method were different among the conformational isomers of 3-CQA. Other isolated compounds such as hydroxycinnamic acid, benzoic acid, coumarin, flavonoid, and lignan compounds also showed high antioxidant activity. Furthermore, a novel chromanone showed a remarkable synergistic effect on ORAC of CQA isomers.

Chapter 2 - Sweet cherries (*Prunus avium* L.) are amongst the most consumed and appreciated fruits worldwide, and are an excellent source of phytochemicals (melatonin, serotonin, carotenoids and phenolic compounds, including flavonoids and anthocyanins) and nutritive substances (organic acids and sugars). The concentrations of these compounds can vary between different sweet cherry cultivars and in different plant parts thereof.

Recently, this fruit has gained more popularity, as there are many scientific studies that evaluate the effects of sweet cherries as health promoters, emphasizing the health benefits of their bioactive compounds, particularly in what concerns their antioxidant, antimicrobial, antidiabetic, anticancer, anti-inflammatory, anti-neurodegenerative and cardiovascular effects, among others.

This chapter will focus on the description of the composition of these fruits, the main factors that influence their profile of bioactive compounds, the different analytical tools to determine the composition, health benefits associated to their consumption, as well as on the recent findings about their potential therapeutic properties.

Chapter 3 - Black cherry (*Prunus serotina* Ehrh.), native to North America, is a tree species appreciated for its valuable wood and ornamental leaves and flowers. Black cherry has been cultivated as an ornamental tree in European gardens, and in Central Europe it has often been planted as a biocenotic admixture under a canopy of Scots pine. In Europe, outside its native North American range, it becomes naturalized and is a highly invasive neophyte.

When the competitiveness and invasiveness of black cherry are compared between its native and exotic range, the duality of the species' behavior can be observed. Its high level of invasiveness in the non-native range can be explained by the 'enemy release' or 'new weapon' hypotheses, as can species-specific growth, reproduction and physiological attributes. In the European occurrence range, growth dynamics of black cherry at a young age are greater than those of co-occurring native tree species. Seedlings allocate more biomass to above-ground organs than to roots when compared with native competitors. Black cherry grows slowly in strong shade and responds to increasing irradiance with faster growth rates. Its large biomass and leaf area enhance light capture and net CO_2 assimilation rates, giving black cherry an advantage over competitors. Both shade and sun leaves of black cherry have shown great plasticity regarding leaf-absorbed energy partitioning. These leaves are rich in nitrogen and may improve soil fertility, but they also contain the toxic cyanogenic glycosides prunasin and amygdalin. These allelochemicals, also found in the seeds and roots of this species, effectively play a protective role against herbivores, but may also inhibit growth of native plants occurring in the same niches. In its non-native range, black cherry is not browsed by ungulates and hosts few pests that are able to feed on tissues containing cyanogenic glycosides. This tree reproduces vegetatively by sprouts and generatively by seeds, which are often dispersed over long distances by birds.

At the seedling and sapling stages, all these traits allow black cherry to outcompete most European tree species. Its strategy of higher biomass investment in above-ground organs, coupled with an increase in light capture and faster net CO_2 assimilation rates, is more important for the success of its invasion outside its native range than any potential allelopathic effects on neighboring native plant species. In European forests, the invasive black cherry has been intensively fought off using mechanical and chemical methods, but without success. The enlargement of its distribution through its invasion of European forests suggests that it may adapt to global climate changes faster than do native species.

Chapter 4 - *Prunus africana* is among the 400 species from the genus *Prunus* that belong to the subfamily Prunoideae of the family Rosaceae, and the only member of the genus with medicinal value that grows in Africa. Propagation of *Prunus africana* is challenging due to the recalcitrant nature of the seed, whose viability is lost after few months post- harvest; the overexploitation due to the unsustainable harvesting which has threatened the species existence and is therefore listed as an endangered species by the Convention on International Trade in Endangered Species (CITES). Thus, endangered species with medicinal and economical value are listed as priority for domestication in many projects in Africa. *In-situ, ex-situ* and *in-vitro* methods have been set for the conservation which ensure mass productivity and sustainability of the species. Besides the various active components found in the bark of *Prunus africana* have been used for decades to treat more than 30 human ailments and several diseases in animals. This bark extract is relatively harmless but should be avoid at higher doses. The fresh cut bark, bruised leaves and crushed seeds have a strong cyanide smell and have shown some poisonous effects. This review examines the classification, toxicity, and techniques used to propagate the endangered *Prunus africana* to improve its medicinal utilization and sustainable conservation.

In: *Prunus*
Editor: William Schneider
ISBN: 978-1-53617-755-8

Chapter 1

PHYTOCHEMICALS IN PRUNES (*PRUNUS DOMESTICA* L.)

Shin-ichi Kayano*[1]*, Hiroe Kikuzaki*[2]*,
***Takahiko Mitani*[3] *and Nobuji Nakatani*[4]**
[1]Department of Nutrition, Faculty of Health Science,
Kio University, Nara, Japan
[2]Department of Food Science and Nutrition, Graduate School of Humanities and Science, Nara Women's University, Nara, Japan
[3]Center for Joint Research and Development,
Wakayama University, Wakayama, Japan
[4]The Open University of Japan, Chiba, Japan

ABSTRACT

Prunes are the dried fruits of some cultivars of *Prunus domestica* L and contain large amounts of phytochemicals. In this study, functional components in prunes such as caffeoylquinic acid (CQA) isomers and quinic acid were quantified by means of HPLC analysis. It has become apparent that prunes contain relatively high amounts of 4-CQA than other fruits, and the content of quinic acid was about 10 times higher than the values in earlier studies. The contribution of CQA isomers to antioxidant activity of prunes was estimated to be 28.4% based on oxygen radical absorbance capacity (ORAC), hence, it was indicated that residual ORAC is dependent on unknown antioxidant components. Total 39 compounds were isolated from prunes and their chemical structures were elucidated on the basis of NMR and MS analyses. Four sesquiterpenoids, three monoterpene derivatives, a chromanone, and a bipyrrole were novel compounds, and two conformational isomers of 3-CQA were also characterized. Antioxidant activities of isolated compounds were evaluated, and each CQA isomer showed high and similar antioxidant activity on Oil Stability Index method, O_2^- scavenging activity, and ORAC. On the other hand, the activities on ORAC and ferric thiocyanate method were different among the conformational isomers of 3-CQA. Other isolated compounds such as hydroxycinnamic acid, benzoic acid, coumarin, flavonoid, and lignan compounds also showed high antioxidant activity. Furthermore, a novel chromanone showed a remarkable synergistic effect on ORAC of CQA isomers.

Keywords: prunes (*Prunus domestica* L.), caffeoylquinic acid, quinic acid, antioxidant activity, HPLC, NMR, MS

INTRODUCTION

Generally, fruits and vegetables are the natural origins of antioxidant components such as flavonoids, flavonolignans, tannins, coumarins, lignans, curcuminoids, phenolic acids, and their derivatives, and show high antioxidant activity (Cao et al., 1998a; Cao et al., 1998b; Potterat, 1997; Wang et al., 1996). Antioxidants in fruits and vegetables are very important on human health because of the large amounts consumed. It is known that phenolic compounds in fruits are potent inhibitors against oxidation of low-density lipoprotein (LDL) *in vitro* (Meyer et al., 1997). Another study has shown that high intake of fruits and vegetables contribute to a significant increase of antioxidant activity of human plasma (Cao et al., 1998a). It is also reported that serum antioxidant capacity was enhanced by consumption of strawberry and spinach in elderly women (Cao et al., 1998b). Antioxidant activity of fruits correlate to their total contents of phenolic compounds (Wang and Lin, 2000; Kalt et al., 1999), and some kinds of fruits such as strawberry, plum, orange, and grapefruit show high antioxidant activity (Wang et al., 1996). In addition, antioxidant activity of prunes was the highest, followed by raisin, blueberry, blackberry, strawberry, raspberry, and other fruits and vegetables based on the oxygen radical absorbance capacity (ORAC) (Agricultural Research Service, 1999).

Prunes are the dried fruits of some cultivars of *Prunus domestica* L. that originated from the Caucasus region in Western Asia. They belong to the Rosaceae family and are one of the species of plums. Generally, plums are the fruits of genus *Prunus*, such as *Prunus domestica*, *P. salicina*, *P. subcordiata*, and *P. insititia* (Pijpers et al., 1986). Among *P. domestica*, so-called prunes are the dried fruits of some cultivars of *P. domestica* that are available for making dried fruits, and these cultivars are called "prune-making plums".

Prunes have been used medicinally in India in combination with other drugs for the treatment of leukorrhoea, irregular menstruation, and debility following miscarriage (Chopra et al., 1956).

In earlier studies, prunes showed considerably higher antioxidant activity than other fruits and vegetables (Agricultural Research Service, 1999). It was reported that high intake of the dietary fibers in the dried fruits of *Prunus domestica* L. lowered low-density lipoprotein (LDL) cholesterol in human plasma (Tinker et al., 1991), as well as plasma and liver lipids in rats (Tinker et al., 1994). Ingestion of prunes also improved ovariectomy-induced hypercholesterolemia in rats (Edralin et al., 2000) and reduced systolic blood pressure of pre-hypertensive subjects (Ahmed et al., 2010). On the bone health, the dried fruits improved the bone mineral density loss in rats (Arjmandi et al., 1999; Deyhim et al., 1999), induced bone formation in postmenopausal women (Arjmandi et al., 2002), and pregnant mice treated with this fruits had newborn mice with high osteogenesis index (Monsefi et al., 2013). On the basis of these and other previous studies, prunes are considered as functional food (Stacewicz-Sapuntzakis et al., 2001).

Concerning the chemical constituents of *P. domestica* L., domesticoside (2-*O*-β-D-glucopyranosyl-4-*O*-methylphloracetophenone) was isolated from the bark of the tree (Nagarajan and Parmar, 1977).

Isosakuranetin, prudomestin, dihydrokaempferide, naringenin, 3,5,7-trihydroxy-8,4′-dimethoxyflavanone, 5,7,4′-trihydroxy-3-methoxy-flavanone, and 3,5,7-trihydroxy-6,4′-dimethoxy-flavanone were isolated from the heartwood (Parmar et al., 1992), and several glycosides of kaempferol and quercetin were detected in the leaves and fruits (Henning and Herrmann, 1980;). Hexanal, (*E*)-2-hexenal, butyl acetate, butyl butyrate and other compounds were found as the aroma compounds of unprocessed prune-making plums (Horvat et al., 1992), and furfural, benzaldehyde, ethyl cinnamate were obtained as flavor components of prunes (Moutounet et al., 1975).

As described above, prunes show high antioxidant activity and contain many kinds of phenolics (Fang et al., 2002), predominant components of which are caffeoylquinic acid (CQA) isomers (Donovan et al., 1998; Raynal et al., 1989). However, there are only a few papers on the quantitative analysis of these isomers in *Prunes domestica* L. In the exocarp and pulp of fresh prune-making plums, the amount of neochlorogenic acid (3-*O*-caffeoylquinic acid, 3-CQA; Figure 1) was about half amount of total phenolics (Raynal et al., 1989). It is also reported that 3-CQA was a major hydroxycinnamate (541 mg/kg) in the fruits of plums followed by chlorogenic acid (5-*O*-caffeoylquinic acid, 5-CQA; Figure 1) at 73 mg/kg, and cryptochlorogenic acid (4-*O*-caffeoylquinic acid, 4-CQA; Figure 1) at 9 mg/kg (Herrmann, 1989). Möller and Herrmann (1983) also detected 88−731 mg/kg of 3-CQA, 15−129 mg/kg of 5-CQA, and 56 mg/kg of 4-CQA in fresh plums. However, as described above, plums consist of some species and both papers did not specify which species of plums were used for analysis, regardless fresh or dried. In a previous study, Donovan et al. (1998) detected 1,300 mg/kg of 3-CQA and 430 mg/kg of 5-CQA in prunes, respectively; however, 4-CQA was not reported in their paper.

Concerning the antioxidant activity of CQA isomers, 5-CQA is recognized to be an antioxidant for human LDL (Rice-Evance et al., 1996; Nardini et al., 1995), a scavenger for reactive oxygen and nitrogen species (Kono et al., 1997), and an inhibitor against formation of conjugated diene from linoleic acid oxidation (Morishita and Kido, 1995). Antioxidant activity of 5-CQA is higher than those of vitamin C and vitamin E based on the Trolox equivalent antioxidant activity (TEAC) (Rice-Evance et al., 1997). Hence, it is expected that other CQA isomers such as 3-CQA and 4-CQA also show high antioxidant activity.

The contribution of CQA isomers to the antioxidant activity of prunes has not been proved in earlier papers. In previous studies, total phenolics of fruits and vegetables were measured for sum of free and conjugated phenolics by hydrolysis method (Vinson et al, 2001; 1998). It is known that the ORAC of fruits and vegetables strongly correlates to its total phenolics (Prior and Cao, 1999a; 1999b). The ORAC and total phenolics of prunes have been evaluated on only soluble fraction; hence, it is required to eliminate the antioxidant activity of whole fruits of prunes.

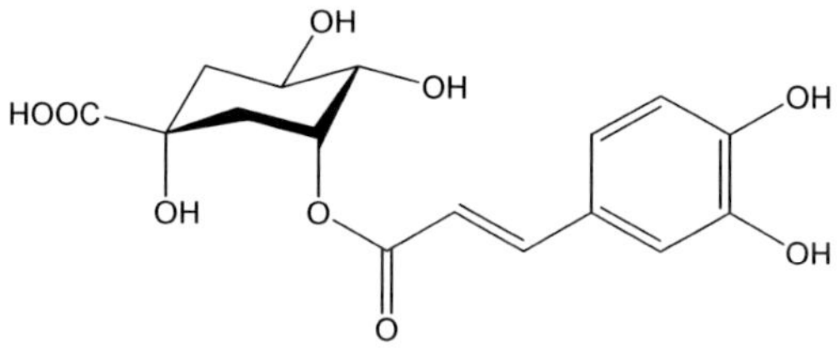

neochlorogenic acid (3-*O*-caffeoylquinic acid, 3-CQA)

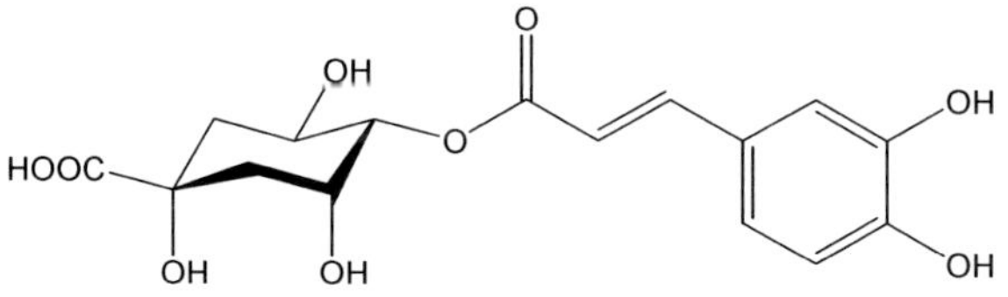

cryptochlorogenic acid (4-*O*-caffeoylquinic acid, 4-CQA)

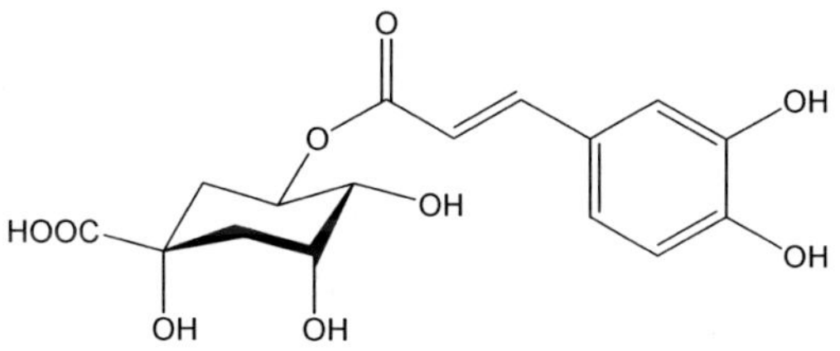

chlorogenic acid (5-*O*-caffeoylquinic acid, 5-CQA)

Figure 1. Structures of caffeoylquinic acid isomers.

Concerning the organic acids in prunes and prune-making plums, it was reported that predominant organic acid is malic acid and the concentration of which is about 0.4 g/100 g; while, the content of quinic acid were lower (0.2 g/100 g) than that of malic acid (De Moura and Dostal, 1965; Fernandez-Flores et al., 1970; Puech and Jouret, 1974). However, Van Gorsel et al. (1992) reported that quinic acid concentration of prune juice, which was water extract of prunes and diluted about 5-fold, was as high as 0.7 g/100 g. Therefore, the content of quinic acid in prunes is unclear in these earlier studies. It is known that quinic acid is converted to hippuric acid via benzoic acid in human (Williams, 1971). Some investigators suggested that consumption of quinic acid acidify human urine by the means of hippuric acid and prevent bacterial overgrowth of urinary tract infections (Schults, 1984). Therefore, if prunes contain large amounts of quinic acid, it might be useful to intake them for treatment and prevention of such urinary diseases.

In the present study on the phytochemicals in prunes, the contents of CQA isomers and organic acid such as quinic acid were elucidated by the means of HPLC analysis. In addition, the contribution of caffeoylquinic acid isomers to the antioxidant activity of prunes was also revealed based on ORAC. Furthermore, isolation, structural elucidation, and antioxidant activity of phytochemicals in prunes are also described.

MATERIALS AND METHODS

Apparatus

^{1}H-, ^{13}C- and 2D NMR (H-H COSY, ^{1}H-^{1}H Correlation Spectroscopy; HMQC, ^{1}H-Detected Multiple Quantum Coherrence Spectrum; HMBC, ^{1}H-Detected Multiple-bond Heteronuclear Multiple Quantum Coherrence Spectrum) spectra were obtained on a Varian Unity plus 500 instrument (^{1}H: 500 MHz, ^{13}C: 125 MHz; Varian Inc., Palo Alto, CA) at 25°C and referenced to the residual proton solvent resonance (CD_3OD at 3.30 ppm for ^{1}H- and 49.0 ppm for ^{13}C-NMR, and acetone-d_6 at 2.05 ppm for ^{1}H- and

29.8 ppm for ^{13}C-NMR). MS analysis was performed on a Hitachi M-1200AP mass spectrometer (Hitachi, Ltd., Tokyo, Japan) with atmospheric pressure chemical ionization (APCI) interface, and HR-MS analysis was carried out on a JMS 700T mass spectrometer (Jeol Ltd., Tokyo, Japan) with a fast atom bombardment (FAB) ionization interface, both of using a methanol (MeOH) solvent. Optical rotations were measured using Jasco P-1030 automatic digital polarimeter (Jasco Co., Tokyo, Japan). HPLC analysis for phenolic compounds was carried out using a Waters 600E multisolvent delivery system equipped with a 717plus autosampler and a 996-photodiode array detector (Waters Co., Milford, MA). HPLC analysis for organic acids was performed using an LC-VP System (Shimadzu Corp., Kyoto, Japan) equipped with a SIL-HTc autosampler (Shimadzu Corp.) and a CDD-10A electroconductivity detector (Shimadzu Corp.). The Omnion Oxidative Stability Instrument (Archer Daniels Midland Co., Decatur, IL) was used for the oxidative stability index (OSI) method described by Akoh (1994). ESR spectra were recorded on a JEOL JES-RE1X spectrometer (JEOL Ltd., Tokyo, Japan) using aqueous quartz flat cell (60 mm × 10 mm × 0.31 mm inner size). The Arvo 1420sx (Wallac Berthold Japan Co., Tokyo, Japan), a type of microplate reader, was used for the measurement of ORAC at an excitation wavelength of 530 nm and emission of 570 nm. Total phenolics assay, proanthocyanidin assay, and ferric thiocyanate method were carried out on a Beckman DU640 spectrophotometer (Beckman Instruments, Inc., Fullerton, CA). IR spectra were run on a Perkin-Elmer 1800 instrument (Perkin-Elmer Inc., Wellesley, MA). Diaion HP-20 (Mitsubishi Chemical Co., Tokyo, Japan), Silica gel 60 (70–230 mesh, E. Merck, Darmstadt, Germany), Sephadex LH-20 (Pharmacia Biotech AB, Uppsala, Sweden) and Chromatorex ODS DM1020T (100–200 mesh, Fuji Silysia Chemical, Tokyo, Japan) were used for column chromatography. Silica gel 60 F_{254} plates (E. Merck, Darmstadt, Germany) and ODS plates (E. Merck, Darmstadt, Germany) were used for thin-layer chromatography (TLC).

Plant Materials

Prunes, which are material for commercial prune extract (concentrated prune juice), and are called "natural condition prune (NC prune)" with moisture levels adjusted to 21% (California Prune Board, 1997), and prune-making plums (*Prunes domestica* L), each of which is the d'Agen cultivar, were imported from California, United States. Frozen cranberry (*Vaccinium macrocarpon*) was provided from Shouei Food Corporation (Tokyo, Japan). Kiwi fruits *(Actinidia deliciosa*), plums (*Prunus salicina*), ume (*Prunus mume*), sweet cherry (*Prunus avium*) and black cherry (*Prunus serotina*) were purchased from a local supermarket in Nishinomiya, Japan.

Standards and Chemicals

Each chemical was obtained as follows, ammonium thiocyanate, L-ascorbic acid, 2,2′-azobis(2-amidinopropane)dihydrochloride (AAPH), caffeic acid, *p*-coumaric acid, ferrous chloride, Folin-Ciocalteu reagent, 5-hydroxymethylfurfural, 7-methoxycoumarin, protocatechuic acid, rutin, and α-tocopherol from Wako Pure Chemical Ins. (Osaka, Japan); chlorogenic acid (5-CQA), quinic acid, 2,6-di-*tert*-butyl-4-methylphenol (butylated hydroxytoluene, BHT), (–)-epicatechin, and 6-hydroxy-2,5,7,8-tetramethylchroman-2-carboxylic acid (Trolox) from Aldrich Chem. Co. (Milwaukee, WI); linoleic acid, malic acid, and methyl linoleate from Tokyo Chemical Industry Co., Ltd. (Tokyo, Japan); delphinidin chloride from Funakoshi Co., Ltd (Tokyo, Japan); diethylenetriamine-*N*, *N*, *N′*, *N″*, *N″*-pentaacetic acid (DTPA) from Dojindo Laboratories (Kumamoto, Japan); 5,5-dimethyl-1-pyrroline-*N*-oxide (DMPO) from LABOTEC Co. (Tokyo, Japan); hypoxanthine (HPX) from Sigma Chemical Co. (St. Louis, MO); β-phycoerythrin (from *Porphyridium cruentum*) from Molecular Probes, Inc. (Eugene, OR); silicon oil from Toshiba Silicon Co., Ltd. (Tokyo, Japan); xanthine oxidase (XOD) from Boehringer Mannheim Co.

(Mannheim, Germany); All commercial standards were of the analytical highest grade.

The compounds, 3-CQA and 4-CQA, were prepared from 5-CQA using the method described by Nagels et al. (1980). The purity of these compounds was higher than 99% for 3-CQA, and 96.5% for 4-CQA by the means of HPLC analysis. The ^{1}H and ^{13}C NMR spectra of 3-CQA are shown in Table 1 and gave good agreement with the literature (Pauli et al., 1999; Pauli et al., 1998).

Table 1. ^{1}H and ^{13}C NMR data of 3-*O*-caffeoylquinic acid (3-CQA), compound 1, and 2

	3-*O*-caffeoylquinic acid (3-CQA)		compound 1		compound 2	
	$^{13}C^{a}$	$^{1}H^{b}$	^{13}C	^{1}H	^{13}C	^{1}H
1	75.4		75.9		75.6	
2ax	36.7	2.20 (*dd*, 3.9, 14.8)	37.4	2.04 (*m*)	37.0	2.09 (*m*)
2eq		2.13 (*m*)		2.04 (*m*)		2.09 (*m*)
3	73.0	5.34 (*ddd*, 3.4, 3.9, 3.9)	72.4	5.40 (*ddd*, 3.1, 4.4, 8.5)	72.7	5.37 (*ddd*, 3.3, 5.1, 5.1)
4	74.8	3.63 (*dd*, 3.4, 8.6)	73.3	3.79 (*dd*, 3.1, 5.7)	74.1	3.70 (*dd*, 3.3, 7.4)
5	68.3	4.14 (*ddd*, 3.9, 8.6, 9.9)	70.1	3.98 (*ddd*, 5.2, 5.7, 5.9)	69.1	4.07 (*ddd*, 4.3, 7.4, 7.7)
6ax	41.5	1.95 (*dd*, 9.9, 13.6)	38.2	1.90 (*dd*, 5.9, 13.9)	40.4	1.93 (*dd*, 7.7, 13.8)
6eq		2.13 (*m*)		2.04 (*m*)		2.09 (*m*)
7	178.3		182.2		180.0	
1′	127.9		128.0		128.0	
2′	115.1	7.04 (*d*, 2.0)	115.1	7.04 (*d*, 2.0)	115.1	7.04 (*d*, 2.2)
3′	146.8		146.8		146.8	
4′	149.4		149.4		149.4	
5′	116.4	6.76 (*d*, 8.3)	116.4	6.76 (*d*, 8.1)	116.4	6.76 (*d*, 8.1)
6′	122.9	6.93 (*dd*, 2.0, 8.3)	122.9	6.94 (*dd*, 2.0, 8.1)	122.9	6.93 (*dd*, 2.2, 8.1)
7′	146.8	7.58 (*d*, 15.9)	146.8	7.58 (*d*, 15.9)	146.8	7.58 (*d*, 15.9)
8′	115.8	6.30 (*d*, 15.9)	115.7	6.30 (*d*, 15.9)	115.8	6.30 (*d*, 15.9)
9′	169.0		168.8		168.9	

[a] ^{13}C NMR were obtained at 125 MHz with CD_3OD (25°C) [b] ^{1}H NMR were obtained at 500 MHz with CD_3OD (25°C).

HPLC Analysis

Phenolic Compounds

For sample preparation of prune fractions (Figure 2), 10 mg of each sample was dissolved in 5 mL of 50% aqueous MeOH. For prune fruits, sample was pitted, cut into small pieces, and then was crushed in an Oster Blender EX (Sunbeam Co., Delray Beach, FL). A portion (10 g) of sample was weighted and homogenized with 100 mL of MeOH in a mixer (Matsushita Electric Co., Osaka, Japan) for 5 min. After homogenization, the mixture was transferred into a flask, added with 50 mL of MeOH, and refluxed for 30 min. The mixture was filtered, and the residue was re-extracted with additional 100 mL of MeOH. This procedure was repeated five times. The combined filtrate was evaporated *in vacuo* to remove the MeOH. The aqueous residue was filled up to 100 mL with water.

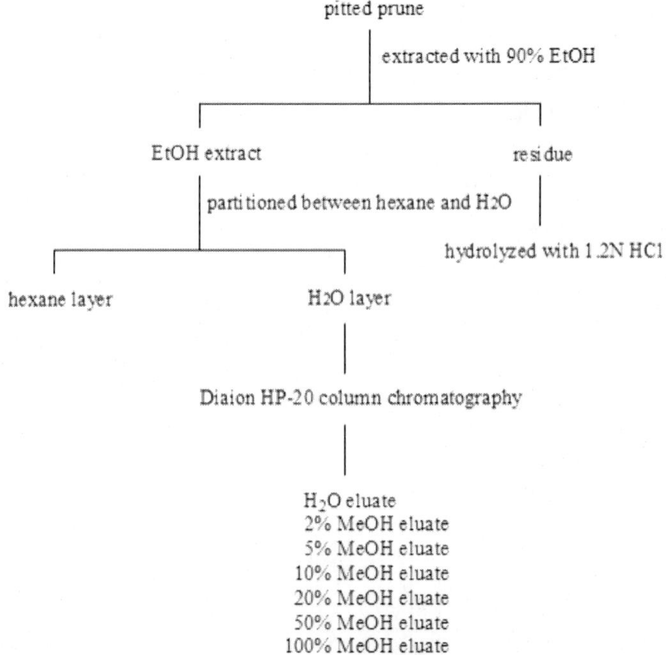

Figure 2. Procedure of extracting and fractionating prunes.

Standards of 3-CQA, 4-CQA, 5-CQA, 5-hydroxymethylfurfural, caffeic acid, *p*-coumaric acid, protocatechuic acid, rutin, (–)-epicatechin, and 7-methocycoumarin were dissolved in 50% MeOH to make a concentration of 1–6 ppm. The sample and standard solutions were filtered through 0.45 μm cellulose syringe-tip filters (Sartorius K. K., Tokyo, Japan) and analyzed by the means of HPLC with the conditions as follows; column, Symmetry C18, 4.6 × 250 mm, 5 μm (Waters Co., Milford, MA); column temp., 40°C; mobile phase, A = 50 mM $NH_4H_2PO_4$ at pH 2.60, B = 80% acetonitrile + 20% A, C = 200 mM *o*-phosphoric acid at pH 1.50; flow rate, 1.0 mL/min; gradient, 0.00 min, %A, = 100.0, 4.00 min, %A = 92.0, %B = 8.0, 10.00 min, %B = 14.0; %C = 86.0, 22.50min, %A = 1.5; %B = 16.5; %C = 82.0, 27.50 min, %B = 21.5, %C = 78.5, 45.00 min, %B = 50.0, %C = 50.0, 47.50 min, %A = 100.0, 55.00 min, %A = 100.0; gradient curve, linear gradient; injection volume, 10 μL; detection, photodiode array (200–600 nm). These conditions refer to the method by Donovan et al. (1998) and are added some modification on gradient. Each peak detected in the sample solutions was identified by comparing the retention time and UV-Vis spectra provided by a photodiode array detector and was quantified by calibration with the standards.

Organic Acids

Prunes were pitted, cut into small pieces, frozen in liquid nitrogen, and crushed with an Oster Blender EX (Sunbeam Co., Delray Beach, FL) to provide analytical sample powder. The edible part of prune-making plums or other fresh fruits was cut into small pieces and was frozen in liquid nitrogen. Immediately after freezing, they were crushed with the blender and were lyophilized to give sample powder. A portion (30 g) of each sample powder (prunes, prune-making plums, or other fruits) was weighed exactly, added with 100 mL of 90% EtOH, and mechanically shaken for 10 min at room temperature. The mixture was centrifuged (3,000 rpm, 10 min), the supernatant was removed, and the precipitate was re-extracted with another portion of 90% EtOH. This procedure was repeated for 19 times. For the quantification of quinic acid and malic acid, each supernatant or combined supernatant was filtered through 0.45 μm cellulose syringe-tip filters (Sartorius K. K., Tokyo, Japan), and analyzed by HPLC with the conditions described afterward.

Another extracting procedure was carried out under refluxing. A portion (5 g) of sample was weighed exactly, added with 50 mL of 50% aqueous MeOH, and the mixture was refluxed for 1 hour. After cooling at room temperature, the mixture was centrifuged, the supernatant was removed, and the precipitate was re-extracted for 3 or 5 times.

Quantification of quinic acid and malic acid in each or combined supernatant was performed by HPLC analysis using an LC-VP System (Shimadzu Corp., Kyoto, Japan) equipped with a SIL-HTc autosampler (Shimadzu Corp.) and a CDD-10A electroconductivity detector (Shimadzu Corp.). The HPLC conditions were as follows: separation conditions, column, Shim-pack SCR-102H (250 mm × 7.8 mm i.d., Shimadzu Corp.), 2-column serial connection, equipped with Shim-pack SCR-102 HG (50 mm × 7.8 mm i.d., Shimadzu Corp.) as guard column; column temperature, 50°C; mobile phase, 2 mM *p*-toluensulfonic acid; flow rate, 0.8 mL/min; detection conditions, reagents, 8 mM Bis-Tris aqueous solution containing 2 mM *p*-toluensulfonic acid and 40 μM EDTA; flow rate, 0.8 mL/min; polarity, +; temperature, 43°C. Standard organic acids were dissolved in 50% MeOH to make a concentration of 6.4–160 ppm for quinic acid and 1.6–40 ppm for malic acid, and these acids in each sample solution was quantified by calibration with the standards.

Total Phenolics Assay

Total phenolics in each fraction was measured using Folin-Ciocalteu reagent with chlorogenic acid as the standard. Each fraction was dissolved in 50% MeOH at 1–50 mg/mL, and 0.5 mL of the sample solution was diluted with 12 mL of H_2O. After addition of 2 mL of 2 M Na_2CO_3 (in H_2O), sample solutions were stored at 20°C for 1 hr. Then to the mixture was added 0.6 mL of Folin-Ciocalteu reagent which was diluted 2-fold with H_2O before use. These prepared samples were further diluted with

H_2O to 20 mL, stored at 20°C for 1 hr. again, and then the optical density of each sample solution at 675 nm was measured using the spectrophotometer. Total phenolics were quantified by calibration using the standard chlorogenic acid solution (0.05–1.5 mg/mL in H_2O).

Proantocyanidin Assay

Proanthocyanidin of EtOH extract residue was measured by colorimetry assay according to the method described in an earlier report (Waterman and Mole, 1994). A portion (50 mg) of EtOH extract residue (freeze-dried) was weighed in a screw-capped tube to which was added 7 mL of 0.6 N HCl in n-butanol (BuOH) with 0.07% of ferrous sulfate. The prepared sample was vortexed for 1 min, heated at 95°C for 1 hr., and filtered. The filtrate was diluted to 20 mL with 0.6 N HCl-BuOH, and then the optical density of sample solution at 550 nm was measured using the spectrophotometer and anthocyanidin was quantified with delphinidin as the standard (0.5–4 mg/100 mL in 0.6 N HCl-BuOH). The analysis was run in duplicate and the data was determined as mean values.

Extraction, Fractionation, and Isolation

Extracting procedures were carried out for total three times. For the 1st extraction, prune fruit (11.24 kg) was pitted, cut into small pieces and homogenized for 5 min with 8 L of 90% aqueous ethanol (EtOH). After filtration of the extract, another 8-L portion of aqueous EtOH was added to the residue. This procedure was repeated four times. The combined extract was evaporated *in vacuo* to remove EtOH, then 4.4 L of hexane was added to the concentrated extract and the mixture was partitioned between hexane and H_2O. Another portion of hexane (4.4 L) was added to the H_2O-soluble fraction and this procedure was repeated five times. The hexane-soluble fractions were combined and evaporated *in vacuo* to give concentrate (77.54 g). The H_2O-soluble fraction was separated by Diaion HP-20 column chromatography using H_2O as an eluting solution followed by elution with MeOH, and each solution was evaporated *in vacuo* to give the H_2O eluate (4770 g) and the MeOH eluate (70.40 g).

A portion of the MeOH eluate (65 g) was re-chromatographed over a column of Sephadex LH-20 gel using 80% aqueous acetone as the mobile phase to give nine fractions (fractions 1–9) by monitoring with silica gel and ODS TLC analysis.

Fraction 5 (7.45 g) was subjected to ODS column chromatography to give fractions 5-1–5-50. Fraction 5-8 (80.9 mg) was subjected on a silica gel column chromatography with ethyl acetate (EtOAc)/MeOH/H_2O/acetic acid (AcOH) (85:10:5:1) followed by CH_2Cl_2/MeOH (15:1) to afford compound 37 (6 mg). Fraction 5-9 (71.4 mg) was re-chromatographed on a column of silica gel with EtOAc/MeOH/H_2O/AcOH (85:10:5:1) followed by ODS (H_2O/MeOH, 75:25) and Sephadex LH-20 (MeOH) column chromatography to afford compound 1 (7 mg), compound 3 (3 mg), and compound 17 (4 mg). Fraction 5-10 (74.6 mg) was subjected to silica gel column chromatography with EtOAc/MeOH/H_2O/AcOH (85:10:5:1) to give compound 5 (11 mg). Fraction 5-11–5-13 were combined (249 mg) and re-chromatographed on a column of silica gel with EtOAc /MeOH/H_2O/ AcOH (77.5:15:7.5:0.7) and Sephadex LH-20 (MeOH) followed by recrystallization (MeOH/Ac) to afford compound 2 (15 mg). Fraction 5-16 (71.8 mg) was subjected to silica gel column chromatography with EtOAc/MeOH/H_2O/AcOH (85:10:5:1) to give compound 15 (4 mg). Fraction 5-18 (39 mg) was re-chromatographed on a column of silica gel with hexane/EtOAc (60:40), ODS with H_2O/CH_3CN (70:30), and silica gel with EtOAc/MeOH/H_2O (70:20:10) to afford compound 18 (4 mg). Fraction 5-27 (372.0 mg) was further purified by ODS column chromatography with CH_3CN/H_2O (9:1) followed by re-chromatography on a column of silica gel with EtOAc/MeOH/H_2O/AcOH (70:20:10) to obtain compound 4 (5 mg) and compound 38 (7 mg).

Fraction 7 (8.06 g) was subjected to silica gel column chromatography with EtOAc/MeOH/H_2O/AcOH (70:20:10:1) to give fraction 7-1–7-5. Fraction 7-1 was further purified by silica gel column chromatography with CH_2Cl_2/Ac (7:3) followed by re-chromatography on a column of

Sephadex LH-20 with MeOH to obtain compound 6 (4 mg), compound 16 (3 mg), and compound 36 (4 mg).

For the 2nd extraction, pitted prune fruits (7.08 kg) were swelled with 14 L of 90% aqueous EtOH followed by homogenizing. After filtration, another 6 L portion of 95% aqueous EtOH was added to the residue and re-homogenized. This procedure was repeated five times. The combined extract was evaporated *in vacuo* to remove EtOH, and then 5 L of n-hexane was added to the residual extract, and the mixture was partitioned between hexane and H_2O. Another portion of hexane (5 L) was added to the H_2O-soluble fraction, and this procedure was repeated five times. The hexane-soluble fractions were combined and evaporated *in vacuo* to give concentrate (74.2 g). The H_2O-soluble fraction was separated by Diaion HP-20 column chromatography using H_2O as an eluting solution followed by elution with MeOH and acetone, and each solution was evaporated *in vacuo* to give the H_2O eluate (2806 g), the MeOH eluate (44.3 g) and the acetone eluate (0.9 g). The MeOH eluate (44.0 g) was re-chromatographed on Sephadex LH-20 gel using 80% aqueous acetone as a mobile phase to give four fractions (fractions 1–4) by monitoring with silica gel and ODS TLC analysis. Fraction 2 (6.77 g) was further subjected to ODS column chromatography eluted with H_2O/CH_3CN (95:5) to afford fifteen fractions (fractions 2-1–2-15). Fraction 2-4 (0.50 g) was re-chromatographed over ODS gel eluted with H_2O/CH_3CN (90:10) to give five fractions (fractions 2-4-1–2-4-5). Each fraction of 2-4-2 and 2-4-3 was purified on ODS column chromatography [H_2O/CH_3CN (90:10)] to give compound 25 (1.7 mg) and 22 (2.8 mg). Fraction 2-4-4 was applied to a Sephadex LH-20 column eluted with 80% aqueous acetone to give compound 23 (5.7 mg). Fraction 2-8 (0.60 g) was re-chromatographed over successive column of ODS [H_2O/CH_3CN (80:20)] and silica gel [EtOAc/MeOH/H_2O (70:20:10)] to afford compound 24 (3.8 mg) and compound 26 (7.0 mg). Fraction 2-7 was subjected to Sephadex LH-20 column chromatography with MeOH to give compound 21 (4.4 mg). Fraction 2-8 was also subjected to Sephadex LH-20 column chromatography with same condition to afford compound 13 (1.5 mg). Fraction 2-10 was re-chromatographed on a column of Sephadex LH-20 with MeOH to give eight fractions (fraction 2-10-1–2-10-

8) and fraction 2-10-4 was further re-chromatographed on a column of ODS with H_2O/CH_3CN (80:20) to obtain fraction 2-4-10-1–2-4-10-19. Fraction 2-10-4-2 was subjected to ODS column chromatography with H_2O/CH_3CN (85:15) to afford compound 35 (11.3 mg). Each of fraction 2-10-4-3, 2-10-4-6, or 2-10-4-12 was re-chromatographed on a column of ODS with H_2O/CH_3CN (80:20) to give compound 31 (11.2 mg) from fraction 2-10-4-3, compound 30 (29.1 mg) from fraction 2-10-4-6, and compound 32 (4.9 mg) from fraction 2-10-4-12. Fraction 2-10-5 was subjected to ODS column chromatography with H_2O/CH_3CN (80:20) to obtain compound 12 (1.3 mg), compound 14 (1.8 mg), and compound 29 (0.9 mg).

Re-chromatography of fraction 2-11 (1.30 g) over Sephadex LH-20 eluted with MeOH gave compound 28 (34.1 mg). Fraction 2-12 (0.25 g) was subjected to a Sephadex LH-20 column eluted with MeOH to give five fractions (fractions 2-12-1–2-12-5). Fraction 2-12-3 was further purified on silica gel column chromatography [CH_2Cl_2/MeOH (80:20)] to afford compound 33 (14.7 mg). Fraction 2-12-4 was purified by successive column chromatography using ODS gel [H_2O/CH_3CN (80:20)] and silica gel [CH_2Cl_2/MeOH (80:20)] to give compound 34 (4.6 mg). Fraction 2-14 (210 mg) was purified on a silica gel column eluted with CH_2Cl_2/MeOH (90:10), and then Sephadex LH-20 column (isopropyl alcohol) to afford compound 27 (2.4 mg).

For the 3rd extraction, prune fruit (1.2 kg) was extracted with 90% MeOH, and the extract was partitioned between hexane and H_2O. The H_2O-soluble fraction was separated by Diaion HP-20 column chromatography using H_2O as an eluting solution followed by elution with 20%, 50% and 100% MeOH. The fractions, 50% MeOH eluate showed high antioxidant activity and low content of CQA isomers. The 50% MeOH eluate was further purified by Sephadex LH-20 column chromatography to give fractions 1–5. Fraction 3 (1.22 g) was subjected to preparative HPLC followed by Diaion HP-20 (H_2O-MeOH) and silica gel (EtOAc/MeOH/H_2O) column chromatography to afford compound 7 (6 mg), compound 8 (5 mg), compound 9 (3 mg), compound 10 (7 mg), compound 11 (7 mg), and compound 20 (2 mg). Fraction 4 (202 mg) was also subjected to preparative HPLC followed by Diaion HP-20 (H_2O-MeOH) and silica gel (EtOAc/MeOH/H2O) column chromatography to obtain compound 19 (2 mg) and compound 39 (6 mg).

The conditions of preparative HPLC were as follows; column, Develosil ODS-HG-5, 25 × 250 mm, 5 μm (Nomura Chemical Co., Aichi, Japan) equipped with Develosil ODS-UG-5, 25 × 50 mm, 5 μm (Nomura Chemical Co., Aichi, Japan) as a guard column; flow rate, 15 mL/min; other conditions are as same as those of analytical HPLC conditions for phenolic compounds described above.

Acetylation

Three mg of prepared 3-CQA, compound 1, or 2 was dissolved in 3 mL of pyridine and heated with hot air. After addition of 3 mL of acetic anhydride, sample solution was heated again and stored overnight at room temperature. Extraction with ethyl acetate from reaction mixture led to give 3-*O*-caffeoylquinic acid pentaacetate (Figure 3) from 3-CQA, and 3-*O*-caffeoylquinide tetraacetate (Figure 3) from 1 and 2. Another acetylation was carried out with acetyl chloride. Two mg of 3-CQA, 1, or 2 was dissolved in 10 mL of acetyl chloride and heated at 60–70 °C for 5–6 hr. to afford 3-*O*-caffeoylquinic acid pentaacetate from 3-CQA, 1, and 2. The NMR data of 3-*O*-caffeoylquinic acid pentaacetate and 3-*O*-caffeoylquinide tetraacetate are as follows:

3-*O*-caffeoylquinic acid pentaacetate: ^{1}H-NMR (500 MHz, CD_3OD): δ 1.99 (1H, *dd*, J = 10.9, 13.8 Hz, H-6ax), 2.03 (3H, *s*, H-4Ac), 2.06 (3H, *s*, H-5Ac), 2.10 (3H, *s*, H-1Ac), 2.32 (3H, *s*, H-4′Ac), 2.33 (3H, *s*, H-3′Ac), 2.42 (1H, *dd*, J = 3.4, 15.9 Hz, H-2ax), 2.65 (1H, *brdd*, H-6ax), 2.78 (1H, *brdd*, H-2eq), 5.08 (1H, *dd*, J = 3.5, 9.9 Hz, H-4), 5.53 (1H, *ddd*, J = 4.3, 9.9,10.9 Hz, H-5), 5.65 (1H, *ddd*, J = 3.4, 3.4, 3.5 Hz, H-3), 6.37 (1H, *d*, J = 16.1 Hz, H-8′), 7.25 (1H, *d*, J = 8.3 Hz, H-5′), 7.38 (1H, *d*, J = 2.0 Hz, H-2′), 7.43 (1H, *dd*, J = 2.0, 8.3 Hz, H-6′), 7.76 (1H, *d*, J = 16.1 Hz, H-7′); ^{13}C-NMR (125 MHz, CD_3OD): δ 20.6 (C-1Ac), 20.6 (C-4Ac), 20.6 (C-5Ac), 20.6 (C-3′Ac), 20.6 (C-4′Ac), 31.9 (C-2), 36.6 (C-6), 66.5 (C-5),

68.2 (C-3), 71.6 (C-4), 78.5 (C-1), 118.4 (C-8′), 122.9 (C-2′), 124.1 (C-5′), 126.5 (C-6′), 132.8 (C-1′), 142.5 (C-3′), 143.8 (C-7′), 143.9 (C-4′), 165.4 (C-9′), 168.0 (C-4′Ac), 168.1 (C-3′Ac), 169.8 (C-1Ac), 170.0 (C-5Ac), 170.2 (C-4Ac), 170.2 (C-7).

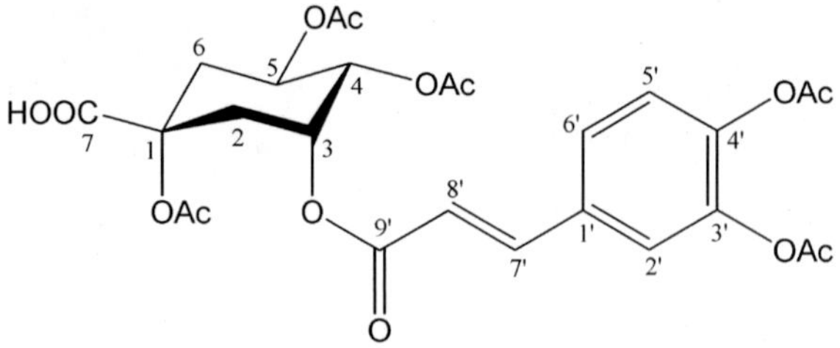

3-*O*-caffeoylquinic acid pentaacetate

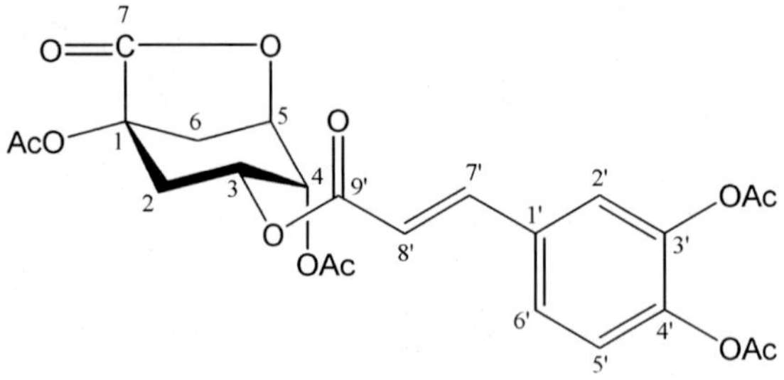

3-*O*-caffeoylquinide tetraacetate

Figure 3. Acetyl derivatives of 3-CQA, compound 1, and 2.

3-*O*-caffeoylquinide tetraacetate: ^{1}H-NMR (500 MHz, CD_3OD): δ 2.16 (6H, *s*, H-1Ac, H-4Ac), 2.31 (3H, *s*, H-4′ Ac), 2.32 (3H, *s*, H-3′Ac), 2.39 (1H, *dd*, J = 10.7, 11.0 Hz, H-2ax), 2.40 (1H, *ddd*, J = 2.2, 7.6, 11.0 Hz, H-2eq), 2.59 (1H, *d*, J = 11.7 Hz, H-6ax), 3.11 (1H, *ddd*, J = 2.2, 6.0, 11.7 Hz, H-6eq), 4.91 (1H, *dd*, J = 5.0, 6.0 Hz, H-5), 5.27 (1H, *ddd*, J = 4.8, 7.6, 10.7 Hz, H-3), 5.56 (1H, *dd*, J = 4.8, 5.0 Hz, H-4), 6.29 (1H, *d*, J = 16.1 Hz, H-8′), 7.23 (1H, *d*, J = 8.5 Hz, H-5′), 7.36, (1H, *d*, J = 2.0 Hz, H-2′), 7.40 (1H, *dd*, J = 2.0, 8.5 Hz, H-6′), 7.62 (1H, *d*, J = 16.1 Hz, H-7′); ^{13}C-NMR (125 MHz, CD_3OD): δ 20.7 (C-3′Ac), 20.7 (C-4′Ac), 20.8 (C-4Ac), 21.1 (C-1Ac), 33.7 (C-2), 34.0 (C-6), 64.9 (C-4), 66.1 (C-3), 73.7 (C-5), 76.2

(C-1), 117.8 (C-8′), 122.8 (C-2′), 124.0 (C-5′), 126.7 (C-6′), 132.9 (C-1′), 142.5 (C-3′), 143.8 (C-4′), 144.3 (C-7′), 164.6 (C-9′), 168.0 (C-4′Ac), 168.1 (C-3′Ac), 169.2 (C-1Ac), 169.2 (C-4Ac), 171.1 (C-7).

Spectral Data of Conformational Isomers of 3-CQA and Novel Compounds

Compound 1

3-*O*-Caffeoylquinic acid (alternative-chair form): pale yellow powder; $[\alpha]^{25}_{D}$ +45.1° (*c*0.49, MeOH); APCI MS, *m/z* 355 $[M+1]^+$; ^{1}H- and ^{13}C-NMR data are shown in Table 1.

Compound 2

3-*O*-Caffeoylquinic acid (skewed boat form): white powder; APCI MS, *m/z* 355 $[M+1]^+$; $[\alpha]^{24}_{D}$ +28.2° (*c*0.45, MeOH); ^{1}H- and ^{13}C-NMR data are shown in Table 1.

Compound 18

4-Amino-4-carboxychroman-2-one: white powder; $[\alpha]^{28}_{D}$ –9.7° (*c*0.42, MeOH); UV (MeOH) λ_{max} (log ε), 253 (3.33), 287 (2.75) nm; IR ν_{max}, 3391 (NH_2), 2925 (CH^2), 1736 (δ-lactone), 1652 (C=O) cm^{-1}; HR-FABMS, *m/z* 206.0482 $[M-H]^-$, calculated for 206.0453 $[C_{10}H_9NO_4-H]^-$; ^{1}H-NMR (500 MHz, CD_3OD) δ 2.64 (1H, *d*, J = 16.0 Hz, H-3), 2.79 (1H, *d*, J = 16.0 Hz, H-3), 6.86 (1H, *dd*, J = 1.0, 7.0 Hz, H-8), 6.99 (1H, *ddd*, J = 1.0, 7.0, 8.0 Hz, H-6), 7.21 (1H, *ddd*, J = 1.0, 7.0, 8.0 Hz, H-7), 7.41 (1H, *dd*, J = 1.0, 7.0 Hz, H-5); ^{13}C-NMR (125 MHz, CD_3OD) δ 43.3 (C-3), 75.5 (C-4), 111.2 (C-8), 123.6 (C-6), 125.1 (C-5), 130.5 (C-7), 133.2 (C-10), 142.8 (C-9), 176.8 (C-11), 181.2 (C-2).

Compound 22

rel-5-(1*R*,5*S*-Dimethyl-3*R*,4*R*,8*S*-trihydroxy-7-oxabicyclo[3,2,1]oct-8-yl)-3-methyl-2*Z*,4*E*-pentadienoic acid: colorless viscous liquid; $[\alpha]^{23}_{D}$ –

15.0° (*c* 0.21, MeOH); UV (MeOH) λ_{max} (log ε), 260.2 (4.21) nm; IR (Nujol) ν_{max}, 3600–2500, 168, 1639, 1602 cm^{-1}; FABMS: *m/z* 389 [M+G-H]$^-$, 297 [M–H]$^-$; HR-FABMS: *m/z* 297.1340 [M–H]$^-$ (calcd for $C_{15}H_{21}O_6$, 297.1338); ^{1}H-NMR (500 MHz, CD_3OD) δ 0.99 (3H, *s* 5′-CH_3), 1.14 (3H, *s*, 1′-CH_3), 1.68 (1H, *dd*, J = 10.5, 14.2 Hz, H-2′ax), 2.03 (1H, *dd*, J = 7.8, 14.2 Hz, H-2′eq), 2.08 (3H, *d*, J = 1.5 Hz, 3-CH_3), 3.48 (1H, *dd*, J = 1.2, 8.5 Hz, H-4′ax), 3.57 (1H, *dd*, J = 1.5, 7.6 Hz, H-6′), 3.76 (1H, *ddd*, J = 7.8, 8.5, 10.5 Hz, H-3′ax), 3.91 (1H, *d*, J = 7.6 Hz, H-6′), 5.76 (1H, *brs*, H-2), 6.31 (1H, *brd*, J = 15.9 Hz, H-5), 7.99 (1H, *dd*, J = 0.7, 15.9 Hz, H-4); ^{13}C-NMR (125 MHz, CD_3OD) δ 12.2 (5′-CH_3), 19.4 (1′-CH_3), 21.2 (3-CH_3), 43.8 (C-2′), 55.0 (C-5′), 72.3 (C-6′), 72.5 (C-3′), 78.1 (C-4′), 82.9 (C-8′), 86.9 (C-1′), 119.4 (C-2), 132.1 (C-4), 134.8 (C-5), 151.3 (C-3), 169.6 (C-1).

Compound 23

rel-5-(1*R*,5*S*-Dimethyl-3*R*,4*R*,8*S*-trihydroxy-7-oxa-6-oxobicyclo[3,2,1] oct-8-yl)-3-methyl-2*Z*,4*E*-pentadienoic acid: colorless viscous liquid; $[\alpha]^{23}_D$ –66.0° (*c* 0.56, MeOH); UV (MeOH) λ_{max} (log ε), 259.0 (4.07) nm; IR (Nujol) ν_{max}, 3600–2500, 1757, 1685, 1654, 1603, 1169 cm^{-1}; FABMS, *m/z* 403 [M+G–H]$^-$, 311 [M–H]$^-$; HR-FABMS: *m/z* 311.1128 [M–H]$^-$ (calcd for $C_{15}H_{19}O_7$, 311.1125); ^{1}H-NMR (500 MHz, CD_3OD) δ 1.18 (3H, *s* 5′-CH_3), 1.33 (3H, *s*, 1′-CH_3), 1.83 (1H, *dd*, J = 9.8, 14.5 Hz, H-2′ax), 2.06 (3H, *d*, J = 1.2 Hz, 3-CH_3), 2.31 (1H, *dd*, J = 7.3, 14.5 Hz, H-2′eq), 3.53 (1H, *d*, J = 9.8 Hz, H-4′ax), 3.58 (1H, *ddd*, J = 7.3, 9.8, 9.8 Hz, H-3′ax), 5.81 (1H, *brs*, H-2), 6.27 (1H, *brd*, J = 16.1 Hz, H-5), 8.02 (1H, *brd*, J = 16.1 Hz, H-4); ^{13}C-NMR (125 MHz, CD_3OD) δ 10.3 (5′-CH_3), 18.5 (1′-CH_3), 21.0 (3-CH_3), 40.5 (C-2′), 59.1 (C-5′), 72.3 (C-3′), 76.6 (C-4′), 82.6 (C-8′), 89.1 (C-1′), 120.7 (C-2), 132.0 (C-5), 133.4 (C-4), 150.5 (C-3), 169.5 (C-1), 178.9 (C-6′).

Compound 24

rel-5-(3*S*,8*S*-Dihydroxy-1*R*,5*S*-dimethyl-7-oxa-6-oxobicyclo[3,2,1]oct-8-yl)-3-methyl-2*Z*,4*E*-pentadienoic acid: colorless solid; $[\alpha]^{25}_D$ –56.7° (*c*0.26, MeOH); UV (MeOH) λ_{max} (log ε), 258.4 (4.27) nm; IR (Nujol)

ν_{max}: 3600–2500, 1757, 1685, 1654, 1602, 1169 cm^{-1}; FABMS: *m/z* 387 [M+G–H]$^-$, 295 [M–H]$^-$; HR-FABMS: *m/z* 295.1196 [M–H]$^-$ (calcd for $C_{15}H_{19}O_6$, 295.1182); ^{1}H-NMR (500 MHz, CD_3OD) δ 1.07 (3H, *s*, 5′-CH_3), 1.34 (3H, *s*, 1′-CH_3), 1.72 (1H, *dd*, J = 11.1, 13.6 Hz, H-4′ax), 1.84 (1H, *dd*, J = 10.1, 14.3 Hz, H-2′ax), 1.89 (1H, *ddd*, J = 1.7, 7.1, 13.6 Hz, H-4′ax), 2.06 (3H, *d*, J = 1.2 Hz, 3-CH_3) 2.23 (1H, *ddd*, J = 1.7, 7.1, 14.3 Hz, H-2′eq), 3.83 (1H, *dddd*, J = 7.1, 7.1, 10.1, 11.1 Hz, H-3′ax), 5.83 (1H, *brs*, H-2), 6.38 (1H, *dd*, J = 0.5, 16.1 Hz, H-5), 7.97 (1H, *dd*, J = 0.7, 16.1 Hz, H-4); ^{13}C-NMR (125 MHz, CD_3OD) δ 14.5 (5′-CH_3), 18.5 (1′-CH_3), 20.9 (3-CH_3), 41.0 (C-4′), 42.3 (C-2′), 53.5 (C-5′), 65.2 (C-3′), 82.8 (C-8′), 89.8 (C-1′), 122.2 (C-2), 131.3 (C-5), 133.5 (C-4), 148.5 (C-3), 171.0 (C-1), 181.1 (C-6′).

Compound 25

rel-5-(3*S*,8*S*-Dihydroxy-1*R*,5*S*-dimethyl-7-oxa-6-oxobicyclo[3,2,1]oct-8-yl)-3-methyl-2*Z*,4*E*-pentadienoic acid 3′-*O*-β-D-glucopyranoside: colorless viscous liquid; $[\alpha]^{22}_D$ –33.2° (*c* 0.09, MeOH); UV (MeOH) λ_{max} (log ε), 258.6 (4.12) nm; FABMS, *m/z* 457 [M–H]$^-$, 277 [M–H–Glc]$^-$; HR-FABMS, *m/z* 457.1716 [M–H]$^-$ (calcd for $C_{21}H_{29}O_{11}$, 457.1710); ^{1}H-NMR (500 MHz, CD_3OD) δ 1.08 (3H, *s*, 5′-CH_3), 1.36 (3H, *s*, 1′-CH_3), 1.88 (1H, *dd*, J=11.2, 13.9 Hz, H-4′ax), 1.94 (1H, *dd*, J = 10.0, 14.2 Hz, H-2′ax), 2.06 (1H, *ddd*, J = 1.2, 7.8, 13.9 Hz, H-4′eq), 2.08 (3H, *d*, J = 1.5 Hz, 3-CH_3), 2.44 (1H, *ddd*, J = 1.2, 7.2, 14.2 Hz, H-2′eq), 3.12 (1H, *dd*, J = 7.8, 9.0 Hz, H-2″), 3.26 (1H, *m*, H-5″), 3.27 (1H, *dd*, J = 9.5, 9.5 Hz, H-4″), 3.31 (1H, *dd*, J = 9.0, 9.5 Hz, H-3″), 3.65 (1H, *dd*, J = 5.4, 12.0 Hz, H-6″α), 3.83 (1H, *dd*, J = 2.2, 12.0 Hz, H-6″β), 3.99 (1H, *dddd*, J = 7.2, 7.8, 10.0, 11.2 Hz, H-3′ax), 4.33 (1H, *d*, J = 7.8 Hz, H-1″), 5.81 (1H, *brs*, H-2), 6.43 (1H, *brd*, J = 15.9 Hz, H-5), 8.01 (1H, *dd*, J = 0.5, 15.9 Hz, H-4); ^{13}C-NMR (125 MHz, CD_3OD) δ 14.5 (5′-CH_3), 18.5 (1′-CH_3), 21.1 (3-CH_3), 39.4 (C-2′), 39.5 (C-4′), 53.5 (C-5′), 62.5 (C-6″), 71.5 (C-4″), 73.0 (C-3′), 75.0 (C-2″), 78.0 (C-3″), 78.0 (C-5″), 82.7 (C-8′), 89.7 (C-1′), 103.2 (C-1″), 120.8 (C-2), 131.9 (C-5), 133.2 (C-18), 150.2 (C-3), 169.8 (C-1), 180.9 (C-6′).

Compound 30

β-D-Glucopyranosyl 7-carboxy-2-methyl-2*E*,4*E*-octadienate: yellow solid; $[\alpha]^{26}_{D}$ –7.3° (*c* 0.35, MeOH); UV (MeOH) λ_{max} (log ε), 268 (3.51) nm; IR (NaCl) ν_{max}, 3359 (OH), 1704 (C=O), 1646 (C=O) cm^{-1}; HR-FABMS: *m/z* 359.1327 $[M–H]^-$ (calcd for $C_{16}H_{23}O_9$, 359.1342); ^1H NMR (500 MHz, CD_3OD) δ 1.16 (3H, *d*, J = 6.8 Hz, 7-CH_3), 1.93 (3H, *d*, J = 1.2 Hz, 2-CH_3), 2.35 (1H, *ddd*, J = 6.8, 8.5, 16.4 Hz, H-6α), 2.53 (1H, *ddd*, J = 6.8, 7.3, 16.4 Hz, H-6β), 2.54 (1H, *ddd*, J = 6.8, 6.8, 8.5 Hz, H-7), 3.35–3.46 (4H, *m*, H-2′, 3′, 4′, 5′) 3.68 (1H, *dd*, J = 4.7, 12.2 Hz, H-6′α), 3.83 (1H, *dd*, J = 1.7, 12.2 Hz, H-6′β), 5.53 (1H, *d*, J = 7.8 Hz, H-1′), 6.12 (1H, *ddd*, J = 6.8, 7.3, 15.0 Hz, H-5), 6.50 (1H, *dd*, J = 11.3, 15.0 Hz, H-4), 7.28 (1H, *dd*, J = 1.2, 11.3 Hz, H-3); ^{13}C NMR (125 MHz, CD_3OD): δ 12.6 (2-CH_3), 17.3 (7-CH_3), 38.2 (C-6), 40.7 (C-7), 62.3 (C-6′), 71.1 (C-4′), 74.0 (C-2′), 78.1 (C-3′), 78.8 (C-5′), 96.0 (C-1′), 126.0 (C-2), 128.9 (C-4), 141.2 (C-3), 142.0 (C-5), 168.6 (C-1), 179.8 (C-8).

Compound 31

β-D-Glucopyranosyl9-carboxy-8-hydroxy-2,7-dimethyl-2*E*,4*E*-nonadienate: pale yellow solid; $[\alpha]^{26}_{D}$ –1.7° (*c* 0.35, MeOH); UV (MeOH) λ_{max} (log ε), 269 (3.64) nm; IR (NaCl) ν_{max}, 3419 (OH), 1700 (C=O), 1637 (C=O) cm^{-1}; HR-FABMS: *m/z* 403.1604 $[M–H]^-$ (calcd for $C_{18}H_{27}O_{10}$, 403.1604); ^1H-NMR (500 MHz, CD_3OD) δ 0.91 (3H, *d*, J = 6.8 Hz, 7-CH_3), 1.69 (1H, *m*, H-7), 1.93 (3H, *d*, J = 1.0Hz, 2-CH_3), 2.08 (1H, *ddd*, J = 8.3, 8.5, 13.9 Hz, H-6α), 2.27 (1H, *dd*, J = 9.0, 15.2 Hz, H-9α), 2.43 (1H, *dd*, J = 3.5, 15.2 Hz, H-9β), 2.49 (1H, *ddd*, J = 4.7, 6.3, 13.9 Hz, H-6β), 3.36–4.46 (4H, *m*, H-2′,3′,4′,5′), 3.67 (1H, *dd*, J = 4.7, 12.1 Hz, H-6′α), 3.77 (1H, *ddd*, J = 3.5, 6.3, 9.0 Hz, H-8), 3.85 (1H, *dd*, J = 1.7, 12.1 Hz, H-6′β), 5.52 (1H, *d*, J = 7.8 Hz, H-1′), 6.18 (1H, *ddd*, J = 6.3, 8.3, 14.9 Hz, H-5), 6.48 (1H, *dd*, J = 11.2,14.9 Hz, H-4), 7.31 (1H, *dd*, J = 1.0, 11.2 Hz, H-3); ^{13}C-NMR (125 MHz, CD_3OD), δ 12.5 (2-CH_3), 15.8 (7-CH_3), 37.2 (C-6), 40.0 (C-7), 41.1 (C-9), 62.3 (C-6′), 71.1 (C-4′), 73.5 (C-8), 74.0 (C-2′), 78.1 (C-3′), 78.8 (C-5′), 96.0 (C-1′), 125.2 (C-2), 128.6 (C-4), 141.6 (C-3), 144.3 (C-5), 168.7 (C-1), 178.0 (C-10).

Compound 32

8-Hydroxy-2,7-dimethyl-2*E*,4*E*-decadienedioic acid 1-β-D-glucopyranyl ester 10-methyl ester: pale yellow solid, $[\alpha]^{26}_{D}$ –8.7° (*c* 0.30, MeOH); UV (MeOH) λ_{max} (log ε), 269 (3.58) nm; IR (NaCl) ν_{max}, 3382 (OH), 1718 (C=O), 1637 (C=O) cm^{-1}; HR-FABMS: *m/z* 417.1747 $[M–H]^-$ (calcd for $C_{19}H_{29}O_{10}$, 417.1761); ^{1}H-NMR (500 MHz, CD_3OD): δ 0.90 (3H, *d*, J = 7.8 Hz, 7-CH_3), 1.70 (1H, *m*, H-7), 1.93 (3H, *d*, J = 0.7 Hz, 2-CH_3), 2.09 (1H, *ddd*, J = 7.9, 9.2, 13.9 Hz, H-6α), 2.38 (1H, *dd*, J = 9.5, 15.1 Hz, H-9α), 2.45 (1H, *ddd,* J = 5.1, 7.3, 13.9 Hz, H-6β), 2.56 (1H, *dd*, J = 3.4, 15.1 Hz, H-9β), 3.36–3.45 (4H, *m*, H-2′, 3′, 4′, 5′), 3.68 (1H, *dd*, J = 5.5, 12.0 Hz, H-6′α), 3.68 (3H, *s*, 10-OCH_3), 3.83 (1H, *dd*, J = 2.2, 12.0 Hz, H-6′β), 3.85 (1H, *ddd*, J = 3.4, 5.1, 9.5 Hz, H-8), 5.53 (1H, *d*, J = 8.0 Hz, H-1′), 6.17 (1H, *ddd*, J = 7.3, 7.9, 15.0 Hz, H-5), 6.48 (1H, *dd*, J = 11.5, 15.0 Hz, H-4), 7.31 (1H, *dd*, J = 0.7, 11.5 Hz, H-3); ^{13}C-NMR (125 MHz, CD_3OD): δ 12.6 (2-CH_3), 15.8 (7-CH_3), 37.1 (C-6), 40.1 (C-9), 40.2 (C-7), 52.2 (10-OCH_3), 62.3 (C-6′), 71.0 (C-4′), 72.9 (C-8), 74.0 (C-2′), 78.0 (C-3′) 78.7 (C-5′), 95.9 (C-1′), 125.3 (C-2), 128.7 (C-4), 141.5 (C-3), 144.0 (C-5), 168.8 (C-1), 174.5 (C-10).

Compound 37

2-(5-Hydroxymethyl-2′,5′-dioxo-2′,3′,4′,5′-tetrahydro-1′*H*-1,3′-bipyrrole)-carbaldehyde: white powder; $[\alpha]^{25}_{D}$ –89.5° (*c* 0.12, MeOH); HR-FABMS, *m/z* 221.0566 $[M–H]^-$, calculated for 221.0563 $[C_{10}H_{10}N_2O_4–H]^-$; UV (MeOH) λ_{max} (log ε), 259 nm (3.76), 294 nm (4.06); IR ν_{max}, 3427 (NH), 1718 (C=O), 1645 (C=O) cm^{-1}; ^{1}H-NMR (500 MHz, acetone-d_6) δ ;2.81 (1H, *dd*, J = 7.1, 17.2 Hz, H-4), 3.18 (1H, *dd*, J = 9.3. 17.2 Hz, H-4), 4.53 (1H, *brs*, 7-OH). 4.73 (2H, *brs*, H-7), 5.79 (1H, *dd*, J = 7.1, 9.3 Hz, H-3′), 6.29 (1H, *d*, J = 4.1 Hz, H-4), 7.10 (1H, *d*, J = 4.1 Hz, H-3), 9.40 (1H, *s*, H-6), 10.19 (1H, *brs*, 1′-NH); ^{13}C-NMR (125 MHz, acetone-d_6) δ 38.2 (C-4′), 56.2 (C-3′), 56.5 (C-7), 110.7 (C-4), 126.2 (C-3), 132.9 (C-2), 145.2 (C-5), 174.9 (C-5′), 175.2 (C-2′), 179.7 (C-6).

Antioxidant Assay

Oil Stability Index Method

Oil stability index (OSI) method was performed according to the method described previously (Akoh, 1994). One μmol of each sample was dissolved in 100 μL of MeOH and added to 5 g of silicon oil that contained 10% of methyl linoleate as lipid substrate. α-Tocopherol and BHT was used for control, and for blank, MeOH alone was added. MeOH was flushed out at 90°C for 30 min with forced aeration after which the conductivity measurement tubes were linked. The lipid substrate is heated with forced aeration, and the effluent air from the substrate going into 50 mL of distilled water contains volatile organic acid swept from the oxidizing oil, which increase the conductivity of the water. The OSI value is defined as the point of maximum change of the rate of oxidation by measuring the conductivity of distilled water. The OSI value of each sample at 90°C was measured in triplicate and the data were given as the mean values ± the standard deviations.

O_2^- Scavenging Activity

Four μmol of each sample were separately dissolved in 20 ml of ultra-pure water. L-Ascorbic acid was used for control. XOD and DTPA solutions were prepared with 200 mM phosphate buffer (pH 7.8), and other solution of chemicals was prepared with ultra-pure water. Fifty μL of 0.4 unit/mL XOD was added to the mixture of sample (50 μL), 2 mM HPX (50 μL), 5.5 mM DTPA (35 μL) and 9.2 M DMPO (15 μL). Exactly 40 seconds after the addition, the ESR spectrum of DMPO-O_2^- was recorded under the following conditions; temperature, 20°C; magnetic field, 335.9 ± 5 mT; power, 8 mW; modulation, 100 kHz; field modulation width, 0.079 mT; sweep time, 2.0 min; receiver gain, 2.5 × 100; time constant, 0.1 sec. All analyses were carried out in triplicate, and the scavenging ratios of O_2^- at 50 μM were calculated as the mean values ± the standard deviations.

Ferric Thiocyanate Method

Antioxidant activity against linoleic acid oxidation was evaluated by the ferric thiocyanate method according to the previous study (Kikuzaki and Nakatani, 1993) with slightly modification. In an amber vial (ϕ = 35, h = 80 mm), 2 mg of each compound was dissolved in 2 mL of water and added with 2 mL of EtOH, 2 mL of 2.51% linoleic acid in EtOH, and 4 mL of phosphate buffer (pH 7.0). After sealing with screw cap, the vial was placed in an oven at 40°C in the dark for 7 weeks. At each test time every week, 0.1 mL of sample solution was added to 9.7 mL 75% EtOH and 0.1 mL of 30% ammonium thiocyanate, and precisely 3 min after addition of 0.1 mL of 0.02 M ferrous chloride in 3.5% hydrochloric acid, the absorbance of the developed color was measured at 500 nm on a spectrophotometer. The measurement was carried out in triplicate and the data were obtained as the mean values ± the standard deviations.

Oxygen Radical Absorbance Capacity (ORAC)

Oxygen radical absorbance capacity (ORAC) was measured according to a method described previously (Cao et al., 1995). This assay is based on the principle that antioxidant compounds delay the decrease of β-phycoerythrin fluorescence induced by AAPH, a peroxyl radical generator.

The test mixture was prepared with 170 μL of 19.6 nM β-phycoerythrin in 75 mM phosphate buffer, pH 7.0, and 10 μL of sample in 75 mM phosphate buffer or acetone at 20μM for pure compounds or 20 μg/mL for crude extract, in microwell plate (96/well, black, Corning Coaster Co., Cambridge, MA). Phosphate buffer alone was used as a blank and 50–400 μM Trolox was used as a control. After 30 min of incubation at 37°C, 20 μL of 300 mM AAPH solution was added to the mixture to initiate the assay and the fluorescence of each well was read every 2 min over a 70 min period at 37°C. The area under the fluorescence curve was calculated and the ORAC value of each sample was expressed as 1 unit for 1 μmol equivalent of Trolox. Each sample was measured in triplicate and the data were determined as the mean values ± the standard deviations. The

ORAC values were measured using an Arvo 1420sx microplate reader at an excitation wavelength of 530 nm and emission wavelength of 570 nm.

Quantification of Functional Components in Prunes by HPLC Analysis

Caffeoylquinic Acid Isomers

Some investigators reported that plums contain small amount of 4-CQA (Herrmann, 1989; Möller and Herrmann, 1983); however, both papers did not specify which species of plum were used for analysis. In another study, Donovan et al. (1998) detected 3-CQA and 5-CQA in prunes, and 4-CQA was not reported in their paper.

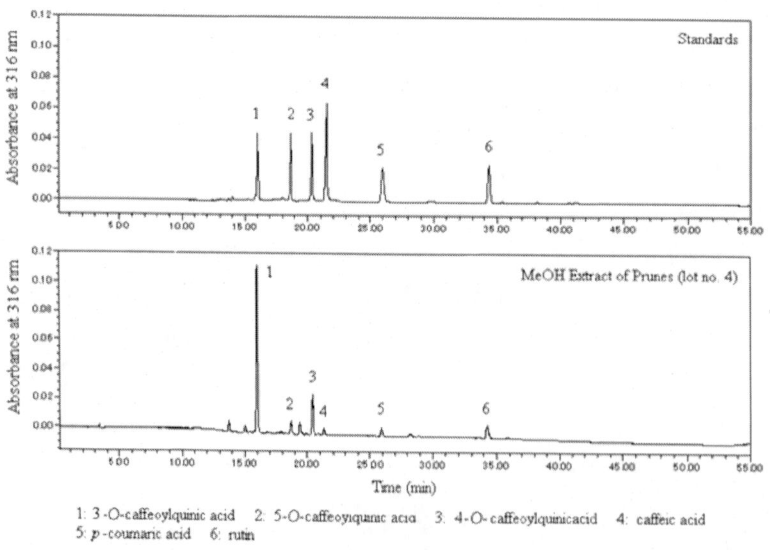

Figure 4. HPLC chromatograms of standard polyphenols and MeOH extract of prunes.

To clarify the contents of CQA isomers in prunes, HPLC analysis was carried out according to the method described in an earlier paper (Donovan et al., 1998). In their HPLC conditions, 3-CQA, 5-CQA, caffeic acid, *p*-

coumaric acid, and rutin were detected in prunes. When the authentic standards were subjected to the HPLC, 4-CQA and caffeic acid appeared at the same retention time and showed a completely overlapped peak. Hence, HPLC conditions were improved with several trials to separate these components, and each peak was resolved to separate completely 4-CQA and caffeic acid as shown in Figure 4.

These phenolics in 4 lots of prunes were measured with these modified HPLC conditions and the analytical data are shown in Table 2. The total contents of phenolic components in prunes were 1,596–1,920 mg/kg in different lots. The predominant compound was 3-CQA, followed by 4-CQA, 5-CQA, and caffeic acid in all lots tested. *p*-Coumaric acid was not detected in lot no.1, and rutin was detected only in lot no.4. In these results, prunes contained 3-CQA, 4-CQA and 5-CQA in the ratio 77.7:18.4:3.9, respectively (calculated based on the mean values of 4 lots of prunes shown in Table 2). This result was apparently different from that of plums in an earlier paper (87:1:12, estimated from Herrmann's study, 1989).

Table 2. Contents of the phenolic components in prunes

	lot no.1 (n=3)	lot no.2 (n=3)	lot no.3 (n=3)	lot no.4 (n=3)
3-*O*-caffeoylquinicacid (3-CQA)	1,228 ± 8[a]	1,233 ± 13	1,485 ± 15	1,372 + 41
4-*O*-caffeoylquinicacid (4-CQA)	288 ± 4	322 ± 6	289 ± 6	351 ± 17
5-*O*-caffeoylquinicacid (5-CQA)	53 ± 2	59 ± 1	77 ± 1	77 ± 2
caffeic acid	26 ± 1	25 ± 1	23 ± 1	28 ± 1
p-coumaric acid *p*-coumaric acid	nd[b]	28 ± 0	22 ± 1	29 ± 1
rutin	nd	nd	nd	65 ± 1
total	1,596 ± 12	1,666 ± 20	1,894 ± 21	1,920 ± 581

[a] Data are expressed as the mean values ± the standard deviations (mg/kg in edible fruit).[b] nd, not detected.

In a previous study of other stone fruits, cherries contained 3-CQA and 3-*O*-*p*-coumaloylquinic acid as major hydroxycinnamates, followed by 5-CQA and other subordinate compounds like 4-CQA. In apricots and peaches, 3-CQA and 5-CQA were predominant and other components were slight. In the pome fruits, apples and pears contained principally 5-CQA and other minor compounds (Möller and Herrmann, 1983).

Concerning these stone and pome fruits, 4-CQA was contained only in cherries as a minor component; therefore, the content of 4-CQA in prunes is fairly high.

It might be supposed that a high amount of 4-CQA was obtained by isomerization among chlorogenic acid isomers, because a previous study happened to find that isomerization of some plant components occurred during the extraction with protic solvent (Nakatani et al., 1991). To confirm whether isomerization of prune components occurred or not during the extraction with MeOH, prunes were extracted with acetone and the amount of chlorogenic acid isomers were measured by HPLC analysis. The content ratio of 3-CQA:4-CQA:5-CQA in acetone extract showed 77.5:18.6:3.9, indicating no difference from that of MeOH extract. On the basis of these results, it is concluded that relatively high amount of 4-CQA in prunes is characteristic.

Organic Acids

Examination of Extracting Conditions

To clarify the contents of organic acids such as malic acid and quinic acid in prunes, HPLC analysis was carried out. In earlier studies, most of investigators measured organic acids by HPLC with a UV detector (Perez et al., 1997; Rodrigez et al., 1992; Tuji et al., 1985); however, quinic acid and malic acid have weak and not specific absorption maximum of UV. Hence, an HPLC system that has a combination of an ion exclusion chromatograph, suitable for separation of organic acids, and an electroconductivity detector capable of selective detection of ionic substances with the post-column pH-buffered method was used in this study. Figure 5 shows typical HPLC chromatograms of standard quinic acid and malic acid, and extract of prune-making plums.

In earlier papers, organic acids such as citric acid, malic acid, or qunic acid in various fruits were extracted with MeOH, aqueous EtOH, or H_2O at cold condition or room temperature (Fernandez-Flores et al, 1970; Perez et al., 1997; Rodriguez et al., 1992; Wills et al., 1986; Tsuji et al., 1985).

According to these studies, extraction of organic acids in prunes and prune-making plums were carried out with 90% EtOH by repeated extraction (19 times) at room temperature and quantified these acids by HPLC analysis with the systems described above. Content of quinic acid in first extract of prunes was 1.1 g/100 g, and total content of quinic acid recovered by 7 times extraction was 3.3 g/100 g. Further extraction (total 19 times) led to increase total amount of quinic acid in prunes to 3.8 g/100 g. On the other hand, content of malic acid in first extract is 0.34 g/100 g, and 0.91 g/100 g in total 19 extracts, respectively (Table 3). Prune-making plums showed similar tendency and total 19 times extractions were required to obtain 0.81 g/100 g of quinic acid and 0.21 g/100 g of malic acid. The yield of organic acids in repeated extraction decreased step by step; however, the amounts of these acids remained in each extract was not negligible. These results suggested that quinic acid and malic acid in prunes and prune-making plums were hard to be completely extracted, and recovery of these acids might be insufficient in these extracting conditions.

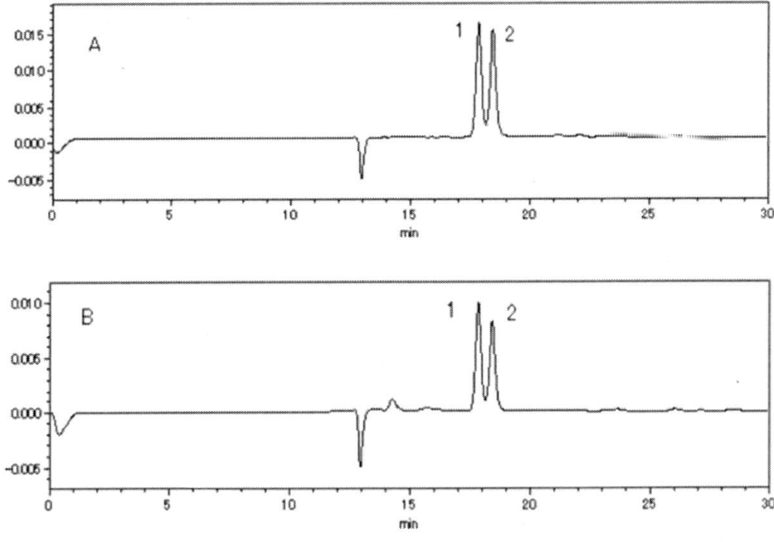

Figure 5. HPLC chromatograms of standard organic acids and extract of prune-making plums.

Table 3. Quinic acid and malic acid contents recovered by repeated extraction with 90% EtOH

sample	organic acids	contents (g/100g) 1st	4th	7th	total of 7 times	total of 19 times
Prunes	quinic acid	1.10	0.32	0.02	3.30	3.80
	malic acid	0.34	0.06	0.04	0.78	0.91
prune-making plums	quinic acid	0.13	0.10	0.06	0.61	0.81
	malic acid	0.06	0.02	0.01	0.16	0.21

Table 4. Quinic acid and malic acid contents recovered under refluxing with 50% MeOH

sample	organic acids	contents (g/100g) 1st	2nd	3rd	4th	5th	total of 5 times
prunes	quinic acid	3.20	0.83	0.23	0.02	nd[a]	4.30
	malic acid	0.90	0.15	0.02	nd	nd	1.10
prune-making plums	quinic acid	0.83	0.25	0.02	nd	nd	1.10
	malic acid	0.25	0.04	nd	nd	nd	0.29

[a] nd, not detected.

Repeated extraction with 90% EtOH at room temperature seemed to be insufficient to extract all of organic acids; hence, extraction with 50% aqueous MeOH under refluxing was carried out. Five times extraction with 50% MeOH under refluxing showed the yield of quinic acid as 3.20 g/100 g in first extract, and 4.30 g/100 g in total extracts (Table 4). Total content of quinic acid was higher than the value obtained with 90% EtOH by 19 times repeated extraction at room temperature (Table 3). Under the same refluxing conditions, prune-making plums also showed higher yield of quinic acid as 0.83 g/100 g in first extract, and 1.1 g/100 g in total 5 times extracts. The content of malic acid showed similar tendency and total contents of this compound was 1.1 g/100 g of prunes and 0.29 g/100 g of prune-making plums, each of which was higher than the value obtained with 90% EtOH by repeated extraction. The content of quinic acid and malic acid in each 4th and 5th extract was slightly; hence, it was suggested that most of organic acids in prunes and prune-making plums were recovered by 3 times extraction in these conditions. It was elucidated that 3 times extraction with 50% MeOH under refluxing was appropriate to extract organic acids in prunes and prune-making plums for quantification of these components.

It is known that prunes contain large amounts of CQA isomers (Donovan et al., 1998; Raynel et al., 1989), which are esters of quinic acid with caffeoyl group. Generally, these esters are easily hydrolyzed to give free quinic acid and caffeic acid, and this cleavage might occur under refluxing condition to show excess quantified value of quinic acid. To confirm this prediction, chlorogenic acid (65 mg) was added to 5.0 g of prune-making plums powder, and quinic acid was extracted with 50% MeOH under refluxing for 3 times. As shown in Table 5, the contents of quinic acid obtained under refluxing were exactly same amounts as 1.10 g/100 g of prune-making plums, whether chlorogenic acid was added or not. Hence, it was clarified that quinic acid esters such as chlorogenic acid in prunes had no influence on quantified value of quinic acid obtained in these extracting conditions.

Table 5. Effect of chlorogenic acid on quantification of quinic acid under refluxing with 50% MeOH

sample	addition of chlorogenic acid (g/100 g)	contents of quinic acid (g/100 g)
prune-making plums	1.30	1.10 ± 0.04[a]
prune-making plums	0	1.10 ± 0.02

[a] Data are presented as the mean values ± the standard deviations.

Quantification of Quinic Acid in Prunes and Other Fruits

It was elucidated that extraction with 50% MeOH under refluxing was suitable for quantification of organic acids in prunes and prune-making plums by HPLC analysis. To compare the contents of quinic acid in prunes and prune-making plums with other materials, several fruits were examined with this method. Typical HPLC chromatograms of cranberry and kiwi fruits are shown in Figure 6. As shown in Table 6, 3 lots of prunes were analyzed and they showed the highest quinic acid contents as 3.70–4.30 g/100g, and prune-making plums also showed higher values as 1.10–1.20 g/100g than other fruits. These values were 5–10 times higher than those of earlier papers, which reported that prunes contain 0.4 g/100 g of quinic acid (Stacewicz-Sapuntzakis et al., 2001; Puech and Jouret, 1974), and the content of prune-making plums is 0.2 g/100 g (De Moure and Dostal, 1965; Fernandez-Flores et al., 1970). The value of quinic acid in prune-making plums in this study was comparable to the content of cranberry, which was previously reported (Hong and Wrolstad, 1986; Wills et al., 1986).

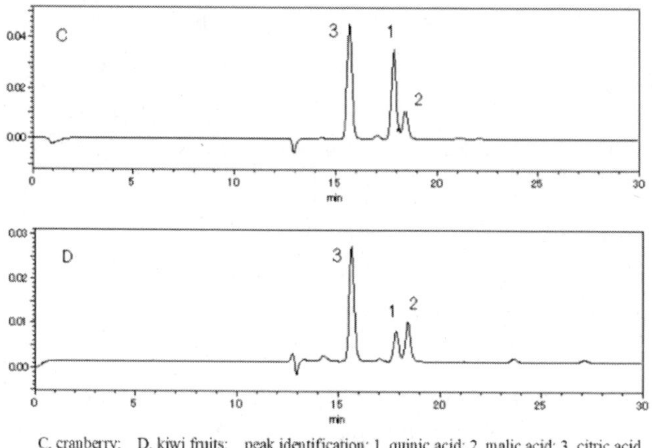

Figure 6. HPLC chromatograms of cranberry and kiwi fruits extracts.

Table 6. Contents of quinic acid in several fruits

sample		contents (g/100 g)[a]	contents in literature (g/100 g)
Prunes	lot no. 1	4.30 ± 0.02	0.4[d,e]
	lot no. 2	3.70 ± 0.03	
	lot no. 3	3.80 ± 0.04	
prune-making plums	lot no. 1	1.10 ± 0.05	0.2[d,e]
	lot no. 2	1.20 ± 0.01	
	lot no. 3	1.20 ± 0.04	
cranberry		0.88 ± 0.02	0.77–1.40[f,g]
kiwi fruits		0.91 ± 0.14	0.56–0.59[h]
plum		0.32 ± 0.02	0.12–0.41[i]
ume		nd[b]	–[c]
sweet cherry		nd	0.05–0.07[i]
black cherry		nd	0–0.01[i]

[a] Data are presented as the mean values ± the standard deviations. [b] nd, not detected. [c] –, no data. [d] Stacewicz-Sapuntzakis et al., 2001. [e] Puech and Jouret, 1974. [f] Hong and Wrolstad, 1986. [g] Wills et al., 1986. [h] Perez et al., 1997. [i] Rodriguez et al., 1992.

Among other fruits, the content of quinic acid in kiwi fruits was relatively high at 0.91 g/100 g followed by cranberry (0.88 g/100 g) and plum (0.32 g/100g). Ume, sweet cherry, and black cherry showed no detectable amounts of quinic acid. These values of analyzed fruits except prunes and prune-making plums were generally in agreement with the literatures (Table 6); hence, it was suggested that quinic acid in prunes and prune-making plums is harder to be extracted than these common fruits.

It was demonstrated that quinic acid in prunes and prune-making plums were higher than the values previously reported, and they seem to be appropriate source of quinic acid for human consumption. It is known that quinic acid is converted to hippuric acid in human to acidify urine (Kinney and Blount, 1979), and the results in this study suggested that prune consumption might be useful for treatment and prevention of urinary tract infections.

Quantitative Evaluation of Antioxidant Components in Prunes

Contribution of Caffeoylquinic Acid Isomers to Antioxidant Activity of Prunes

The contribution of CQA isomers to the antioxidant activity of prunes has not been proved in earlier papers. To clarify this point, ORAC and the contents of CQA isomers in EtOH extract of prunes were measured and the contribution was estimated. Prune fruits (205 g) were pitted (175 g of edible portion), cut into small pieces, and homogenized with 90% aqueous EtOH. After filtration of the extract, another portion of aqueous EtOH was added to the residue and re-extracted. The combined extract was evaporated *in vacuo* in order to remove EtOH, followed by dissolution with H_2O, then hexane was added, and the mixture was partitioned between hexane and H_2O. After separation of the hexane layer, another portion of hexane was added to the H_2O layer and this procedure was repeated five times. The hexane layers were combined and evaporated *in vacuo* to produce a concentrate. The H_2O layer was separated using Diaion HP-20 column chromatography employing H_2O as an eluting solution, followed by elution with 2%, 5%, 10% 20%, 50%, and 100% MeOH successively, and each solution was evaporated *in vacuo* to give the H_2O, and 2%, 5%, 10%, 20%, 50%, and 100% MeOH eluates (Figure 2).

Table 7. Yields, total phenolics, oxygen radical absorbance capacity (ORAC), contribution of CQA[a] isomers, and proanthocyanidin of prune extract and residue

	yield[b]	total phenolics[c]	CQA isomers[d]	total ORAC[e]	ORAC of CQA isomers[f]	contribution of CQA isomers[g]	proanthocyanidin[h]
	(g)	(mg)	(mg)	(units)	(units)	(%)	(mg)
EtOH extract	122.0	3380	359	12800	3660	28.4	
hexane layer	1.4	144	nd[i]	312		0	
H_2O layer	119.0	3140	342	11600	3480	30.0	
residue	33.7[j]	3780[k]		21700[l]			108

[a] CQA, caffeoylquinic acid. [b] Recovered weight of each fraction from 205 g of prunes. [c] Total phenolics are expressed as chlorogenic acid equivalent. [d] Sum of the contents of CQA isomers in each fraction determined by HPLC analysis. [e] Total ORAC values are calculated as ORAC of each fraction (units/mg; determined by ORAC assay) × yield (mg). [f] Calculated as CQA isomers (mg) × ORAC of chlorogenic acid (3.6 units/μmol; determined by ORAC assay). [g] Calculated as ORAC of CQA isomers × total $ORAC^{-1}$ × 100. [h] Proanthocyanidin was expressed as delphinidin equivalent. [i] nd, not detected. [j] weight of freeze-dried residue. [k] total phenolics of hydrolyzed residue. [l] total ORAC of hydrolyzed residue.

Total phenolics and the ORAC of each fraction were measured, and CQA isomers in these fractions were quantified by the means of HPLC analysis. The EtOH extract (122.0 g), the recovery of which was about 70% of the pitted prune, was separated into a hexane layer and a H_2O layer, and the yields of these fractions were 1.4 g and 119.0 g, respectively (Table 7). Both of total phenolics and ORAC value of the H_2O layer was 20–40 times higher than those of the hexane layer (Table 7), and it was predicted that almost all antioxidant components in prunes are also entirely in the H_2O layer with sugars and organic acids. The total content of CQA isomers in EtOH extract was 359 mg measured by HPLC analysis, and the ORAC of these isomers was 3660 units. However, the total ORAC of EtOH extract was 12800 units, and the contribution of CQA isomers to the ORAC of this fraction was as low as 28.4% (Table 7). Furthermore, the contribution of these isomers to the ORAC of the H_2O layer was also revealed to be as low as 30.0%, hence, it was suggested that unknown antioxidants exist in the H_2O layer and total ORAC of them are higher than those of CQA isomers.

Properties of Antioxidant Components in H_2O Layer Fractions

The H_2O layer, which showed high total phenolics and total ORAC (Table 7), was further separated into H_2O, and 2%, 5%, 10%, 20%, 50%, and 100% MeOH eluates by the means of Diaion HP-20 column chromatography (Figure 2). The yield of H_2O eluate was as high as 110 g (Table 8), and this fraction showed the highest total phenolics (1200 mg) and total ORAC (4720 units) among the separated fractions. Each of these values corresponded to about 40% of those of the H_2O layer (calculated based on the values shown in Table 8). In the HPLC analysis, three major peaks were detected in the H_2O eluate (Figure 7), two of which were identified as 5-hydroxymethylfurfural and 3-CQA. The content of CQA isomers in the H_2O eluate was 64.6 mg, and the contribution of these isomers to ORAC was relatively low at 13.9% (Table 8). In this study, 5-hydroxymethylfurfural, sugars, and organic acids were negative on total phenolics and ORAC assay (data not shown); hence, it was speculated that unknown antioxidant components exist in this fraction. Some of other minor peaks were identified as protocatechuic acid and 7-methoxycoumarin, and it is presumed that other major and unidentified peaks suggest the presence of high polar antioxidant components in this fraction.

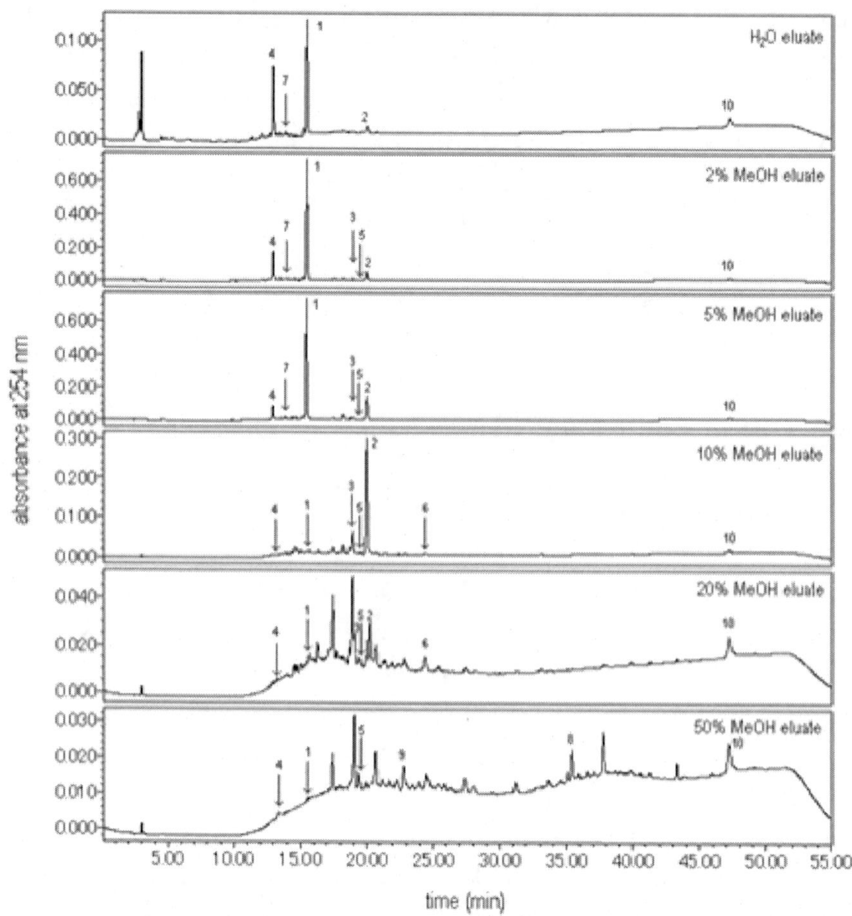

1. 3-*O*-caffeoylquinic acid, 2. 4-*O*-caffeoylquinic acid, 3. 5-*O*-caffeoylquinic acid, 4. 5-hydroxymethylfurfural, 5. caffeic acid, 6. *p*-coumaric acid, 7. protocatechuic acid, 8. rutin, 9. (-)-epicatechin, 10. 7-methoxycoumarin

Figure 7. HPLC chromatogram of prune fractions.

The other fractions (2–100% MeOH eluates) had 0.17–0.46 g yields and showed about 60% of the total ORAC of the H_2O layer. Concerning the 2% and 5% MeOH eluates, their total ORACs were as high as 1430–1500 units; however, total phenolics were relatively small at 201–209 mg (Table 8). In the HPLC analysis, about 60% of the CQA isomers of the H_2O layer existed in these eluates (calculated based on the values shown in Table 8), and the contribution of these isomers to ORAC were as high as 70.2–70.7%. Each major peak detected in these fractions by HPLC analysis was identified as 3-CQA, 5-hydroxymethylfurfural, and 4-CQA, respectively, and caffeic acid, protocatechuic acid, and 7-methoxycoumarin were also identified as minor components. Concerning the 10% and 20% MeOH eluates, the contribution of CQA isomers to the ORAC of these fractions were relatively low at 7.9–38.3%, and each of the total phenolics and total ORAC was also low at 127–150 mg and 880– 928 units, respectively. In the HPLC analysis, several unknown peaks were detected, and some of them were identified as caffeic acid, *p*-coumaric acid, and 7-methoxycoumarin.

Table 8. Yields, total phenolics, oxygen radical absorbance capacity (ORAC), and contribution of CQA[a] isomers of H_2O layer fractions

	yield[b]	total phenolics[c]	content of CQA isomers	total ORAC[d]	ORAC of CQA isomers[e]	contribution of CQA isomers[f]
	(g)	(mg)	(mg)	(units)	(units)	(%)
H_2O layer	119	3140	342.0	11600	3480	30.0
H_2O eluate	110	1200	64.6	4720 (40.7)[g]	656 (18.9)[h]	13.9
2% MeOH eluate	0.29	209	99.3	1430 (12.3)	1010 (29.0)	70.7
5% MeOH eluate	0.27	201	103.8	1500 (12.9)	1060 (30.5)	70.2
10% MeOH eluate	0.17	127	35.0	928 (8.0)	356 (10.2)	38.3
20% MeOH eluate	0.20	150	6.9	880 (7.6)	70 (2.00)	7.9
50% MeOH eluate	0.46	531	0.1	1860 (16.0)	1 (0.02)	0.1
100% MeOH eluate	0.19	229	n.d	608 (5.2)	0	0

[a] CQA, caffeoylquinic acid. [b] Recovered weight of each fraction from 205 g of prune. [c] Total phenolics are expressed as chlorogenic acid equivalent. [d] Total ORAC values are calculated as ORAC of each fraction (units/mg; determined by ORAC assay) × yield (mg). [e] Calculated as amounts of CQA isomers (mg, shown in Table 1) × ORAC of chlorogenic acid (3.6 units/μmol; determined by ORAC assay). [f] Calculated as ORAC of CQA isomers × total $ORAC^{-1}$ × 100. [g] Recovery ratio (%) of ORAC to that of H_2O layer are expressed in parenthesis. [h] Recovery ratio (%) of ORAC of CQA isomers to that of H_2O layer are expressed in parenthesis.

Concerning the 50% MeOH eluate, the contribution of CQA isomers to the ORAC was only small; however, the eluate showed relatively high total ORAC of 1860 units corresponding to 16.0% of that of the H_2O layer (calculated based on the values shown in Table 8), and this fraction also indicated relatively high total phenolics at 531 mg. Many peaks were detected by HPLC analysis (Figure 7), and some of them were identified as 5-hydroxymethylfurfural, caffeic acid, rutin, (–)-epicatechin, and 7-methoxycoumarin, respectively (Figure 8). Caffeic acid, rutin, and (–)-epicatechin are known as antioxidant components (Potterat, 1997; Rice-Evans et al., 1996; Prior and Cao, 1999a) and are considered to take part in the ORAC of this fraction, and it is predicted that the detections of other unidentified peaks indicate the presence of unknown antioxidant components in this fraction. The 100% MeOH eluate showed 608 units of total ORAC and no contribution of caffeoylquinic acid isomers, and this fraction also might contain unknown antioxidant components.

5-hydroxymethylfurfural

caffeic acid

p-coumaric acid

protocatechuic acid

rutin

(-)-epicatechin

7-methoxycoumarin

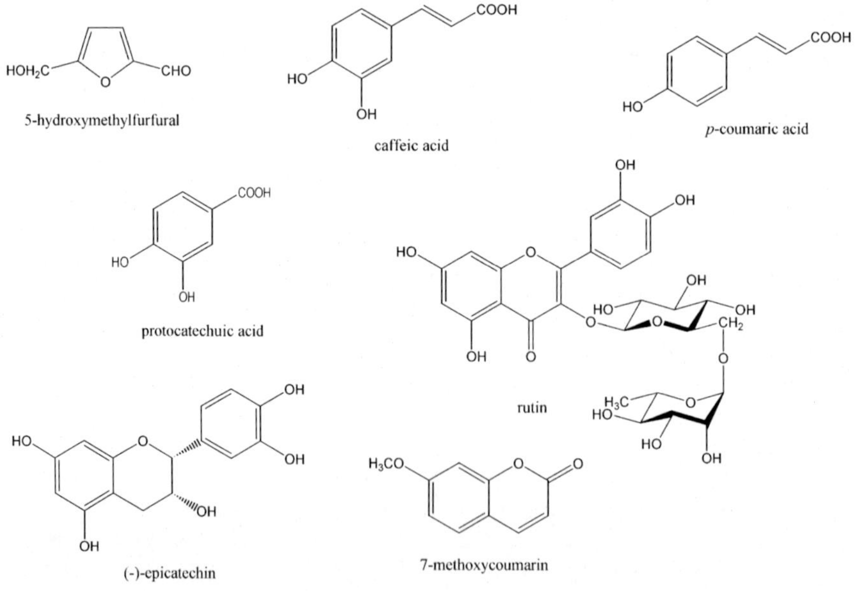

Figure 8. Detected phytochemicals in prune fractions.

Antioxidant Activity of Hydrolyzed Residue

In earlier studies concerning prunes, antioxidant compounds were characterized only in soluble fractions by the means of HPLC analysis and the evaluation of insoluble antioxidants in prunes were insufficient. Therefore, the EtOH extract residue of prunes was hydrolyzed and total phenolics and ORAC were evaluated to clarify the presence of unknown antioxidants in prunes. Hydrolyzed residue showed high total phenolics at 3780 mg which was almost the same amount as that of the EtOH extract (Table 7). This fraction also indicated a high total ORAC (21700 units), and it was 1.4 times higher than that of the EtOH extract. It is presumed that almost all of the soluble antioxidants are entirely in the EtOH extract; hence, large amounts of conjugated compounds such as insoluble tannin or proanthocyanidin exist in the residue and were extracted as lower molecule compounds obtained by hydrolysis. In order to estimate the contents of insoluble antioxidants in the EtOH extract residue, proanthocyanidin assay was carried out. The contents of proanthocyanidin was 108 mg (Table 7), and these results is considered to indicate that proanthocyanidin take part in antioxidant activity of the residue. However, contents of proanthocyanidin are apparently smaller than total phenolics of hydrolyzed residue, and it is predicted that other conjugated antioxidant still exist in prunes.

In a previous report concerning the prune components, proanthocyanidin was detected at 0.79 mg/g in dried pulp by the means of colorimetry assay (Rice-Evans et al., 1996), and the value is similar to the result in this study (0.78 mg/g, calculated based on the value of Table 7). On the other hand, no proanthocyanidin was detected by LC/MS/MS analysis in a previous study, which analyzed MeOH soluble fraction of prunes (Fang et al., 2002). The available studies regarding tannin or proanthocyanidin are insufficient, and thus characterization of conjugated antioxidants in prunes is required.

Isolation and Structural Elucidation of Phytochemicals in Prunes

Extraction, Fractionation, and Isolation

To clarify unknown antioxidants in prunes, extraction, isolation, and structural elucidation of prune components were carried out. Extracting procedures were made for total three times. For the 1st extraction, prune fruit was extracted with 90% aqueous EtOH, and the extract was separated into the hexane-soluble and the H_2O-soluble fractions. The H_2O-soluble fraction was separated by Diaion HP-20 column chromatography into the H_2O eluate and the MeOH eluate. The MeOH eluate was re-chromatographed over a Sephadex LH-20 gel column with 80% aqueous acetone to give fractions 1–9. Further purification of fraction 5 and 7 on various column chromatography led to isolation of six hydroxycinnamic acids (compound **1–6**), three benzoic acids (compound **15–17**), a chroman (compound **18**), and three other compounds (compound **36–38**).

For the 2nd extraction, prune fruit was extracted with 90% aqueous EtOH, and the extract was separated into the hexane-soluble and the H_2O-soluble fractions. The H_2O-soluble fraction was separated by Diaion HP-20 column chromatography into the H_2O, MeOH, and acetone eluate. The MeOH eluate was further chromatographed on a column of Sephadex LH-20 gel with 80% aqueous acetone to obtain fractions 1–4. Various column chromatography of fraction 2 afford three hydroxycinnamic acids (compound **12–14**), a coumarin (compound **21**), seven abscisic acid related compounds (compound **22–28**), a monoterpene and three monoterpene derivatives (compound **29–32**), and three lignans (**33–35**).

For the 3rd extraction, prune fruit was extracted with 90% MeOH, and the extract was partitioned between hexane and H_2O. The H_2O-soluble fraction was separated by Diaion HP-20 column chromatography using H_2O as an eluting solution followed by elution with 20%, 50% and 100% MeOH. The fractions, 50% MeOH eluate were further purified by Sephadex LH-20 column chromatography followed by preparative HPLC and silica gel column chromatography to give five hydroxycinnamic acids (compound **7–11**), two coumarins (compound **19**, **20**), and a flavonoid (compound **39**).

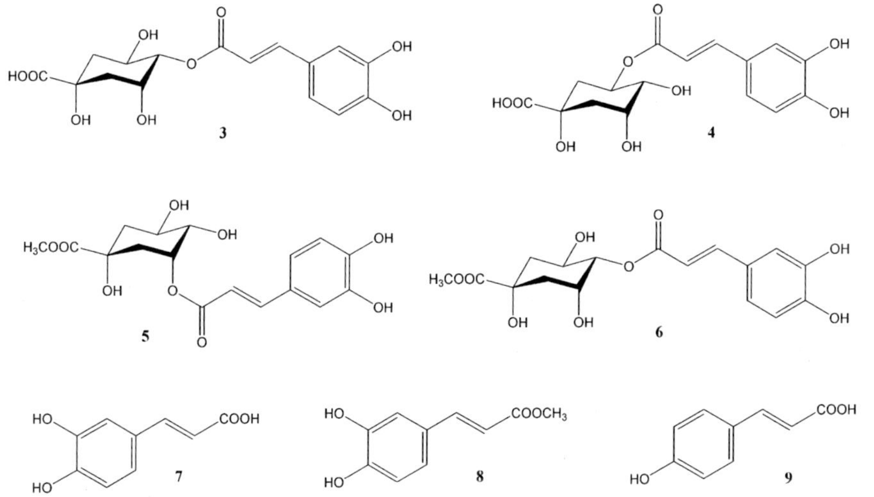

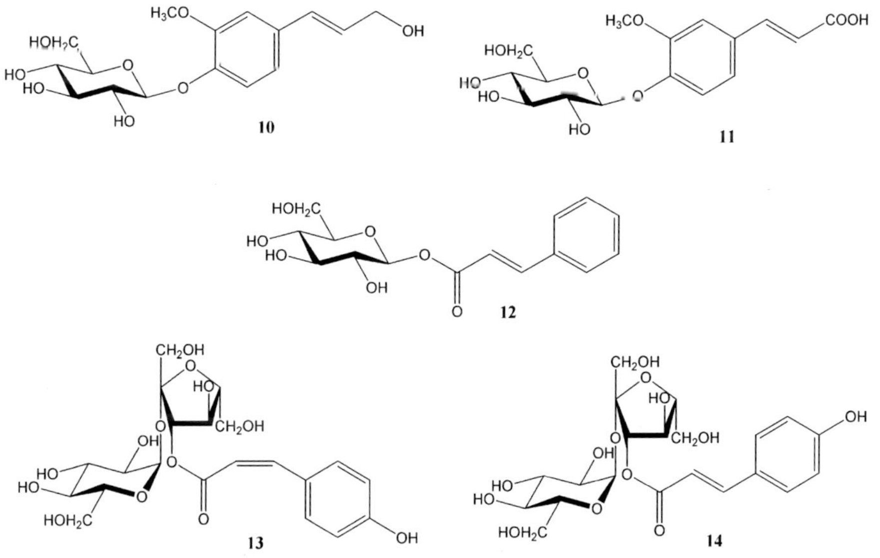

Figure 9(A,B) Isolated hydroxycinnamic acids.

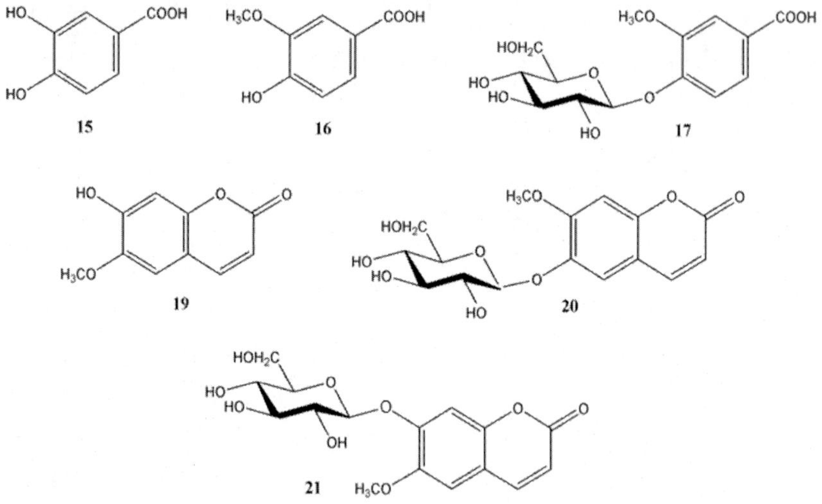

Figure 10. Isolated benzoic acids and coumarins.

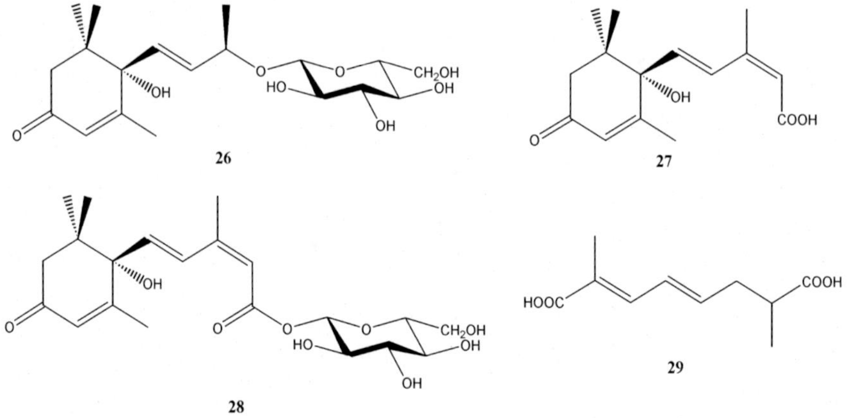

Figure 11. Isolated abscisic acid related compounds and monoterpen.

Finally, fourteen hydroxycinnamic acids (**1−14**), three benzoic acids (**15−17**), a chroman (**18**) and three coumarins (**19−21**), seven abscisic acid related compounds (**22−28**), a monoterpene and three monoterpene derivatives (**29–32**), three lignans (**33−35**), and four other compounds (**36−39**) were obtained in these extraction and isolation. The chemical structure of these isolated compounds were elucidated on the basis of the NMR and MS analyses and determined to be 4-*O* caffeoylquinic acid (**3**),

5-*O*-caffeoylquinic acid (**4**), 3-*O*-caffeoylqunic acid methyl ester (**5**), 4-*O*-caffeoylquinic acid methyl ester (**6**), caffeic acid (**7**), caffeic acid methyl

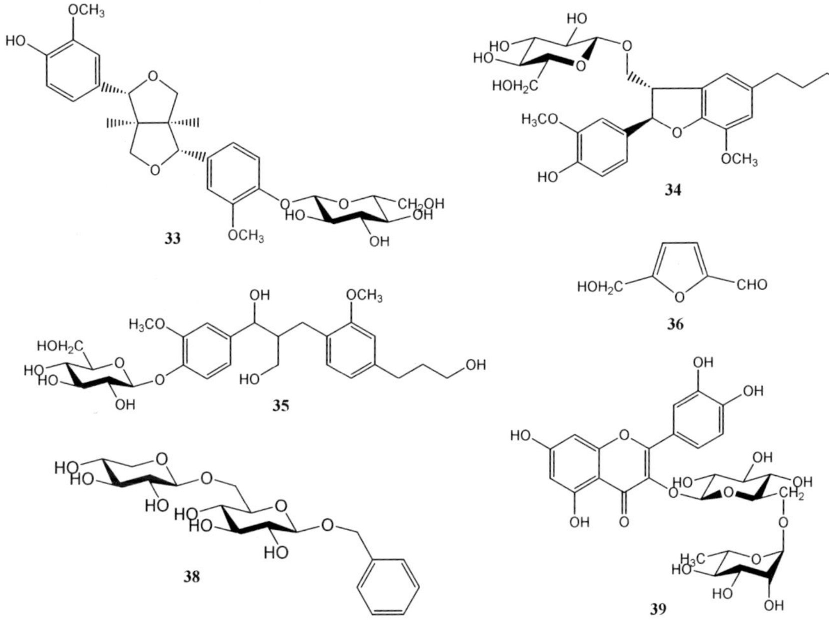

Figure 12. Isolated lignans and other compounds.

ester (**8**), *p*-coumaric acid (**9**) (Figure 9A), coniferin (**10**), ferulic acid (**11**), cinnamic acid β-D-glucopyranoside (**12**), (3-*O-cis-p*-coumaroyl-β-D-fructofuranosyl)-2→1)-α-D-glucopyranoside (**13**) (3-*O-trans-p*-coumaroyl-β-D-fructofuranosyl)-2→1)-α-D-glucopyranoside (**14**) (Figure 9B), protocatechuic acid (**15**), vanillic acid (**16**), vanillic acid β-D-glucopyranoside (**17**), scopoletin (**19**), magnolioside (**20**), scopolin (**21**) (Figure 10), (6*S*,9*R*)-roseoside (**26**), abscisic acid (**27**), β-D-glucopyranosyl abscisate (**28**), 2,7-dimethyl-2*E*,4*E*-octadiene-1,8-dicarboxylic acid (**29**) (Figure 11), (+)-pinoresinol *O*-β-D-glucopyranoside (**33**), 9-(β-D-glucopyranosyl)-7-(4-hydroxy-3-methoxyphenyl)-1-(3-hydroxypropyl)-3-methoxy-(7*R*,8*S*)-dihydrrobenzofuran (**34**), 1*S*-(4-β-D-glucopyraosyl-3-methoxyphenyl)-2*R*-[4-(3-hydroxypropyl)-2-methoxyphenyl]-1,3-

propandiol (**35**), 5-hydroxymethylfurfural (**36**), β-primeveroside (**38**), and rutin (**39**) (Figure 12), respectively.

Compound **3**, **5**, **6**, **8**, **11–17**, **19**, **20**, **26–29**, **33**, **34**, and **35** were isolated from prunes (*Prunus domestica* L.) for the first time.

Structural Elucidation of Conformational Isomers of 3-CQA and Novel Compounds

Conformational Isomers of 3-CQA

Compound **1** was obtained as a white powder and exhibited an $[M+1]^+$ peak at *m/z* 355, which was indicative of the molecular weight of 354 corresponding to caffeoylquinic acid. The ^{13}C-NMR data revealed the presence of a quinic acid moiety characterized with two methylenes (δ_C 37.4 and 38.2), three oxymethines (δ_C 70.1, 72.4, and 73.3), one quaternary carbon (δ_C 75.9) and one carboxyl group (δ_C 178.3) as well as a caffeoyl moiety. In the ^{1}H-NMR spectrum, the downfield shifted signal (*ddd*) of oxymethine proton of quinic acid moiety (δ_H 5.40) suggested that hydroxyl group at 3- or 5-position was esterified with caffeoyl group. However, the signals of ^{1}H and ^{13}C NMR spectra of **1** different from those of authentic 3- and 5-CQA, on the oxymethine and methylene of quinic acid region. In addition, optical rotations of **1** ($[\alpha]^{25}_D$, +45.1°, *c*0.49, MeOH) was also different from that of 3-CQA ($[\alpha]^{23}_D$, +11.7°, *c*0.49, MeOH) or 5-CQA ($[\alpha]^{24}_D$, −37.6°, *c*0.50, MeOH). Furthermore, acetylation of 3-CQA with acetic anhydride and pyridine gave 3-*O*-caffeoylquinic acid pentaacetate; while, acetylation of **1** with same condition afforded a caffeoylquinide tetraacetate (Figure 3), bearing a γ-lactone ring formed with carboxyl group (C-1) and hydroxyl group (C-5) in the quinic acid moiety. However, acetylation of 3-CQA, or **1** with acetyl chloride resulted in the formation of 3-*O*-caffeoylquinic acid pentaacetate; hence, it was presumed that **1** is a conformational isomer of 3-CQA on quinic acid molecule.

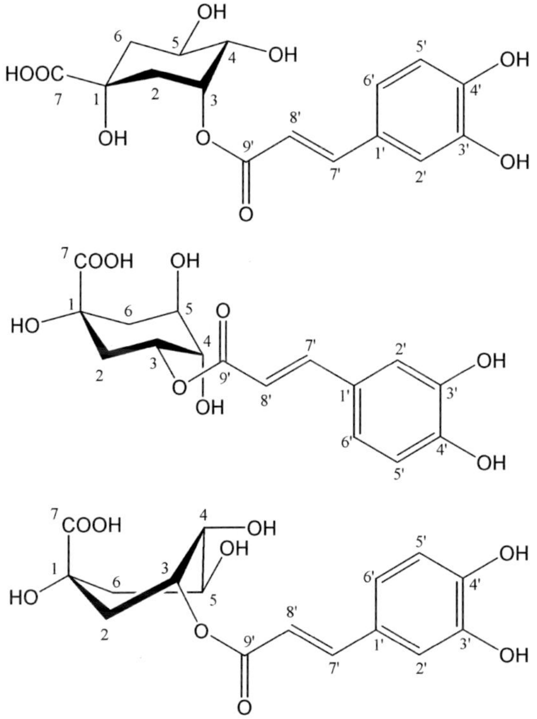

3-*O*-caffeoylquinic acid (chair form)

compound **1** (alternative-chair form)

compound **2** (skewed boat form)

Figure 13. Stereochemical conformation of 3-CQA, compound **1**, and **2.**

To clarify the stereochemistry of **1**, NMR techniques were carried out. In the ^{1}H NMR analysis (Table 1), an oxymethine proton (δ_H 5.40, H-3) showed a large vicinal coupling constant (8.5 Hz) with a neighboring methylene proton (H-2α), and it was assigned that this proton (H-3) is axial in orientation. In addition, the proton at the 4-position (δ_H 3.79) must be equatorial because of its small coupling constant (3.1 Hz) with the axial proton at the 3-position. In these orientations, stereochemistry of quinic acid moiety of **1** should be either boat form, proton at the 5-position of which is axial, or alternative-chair form, proton at the 5-position is equatorial. This oxymethine proton (δ_H 3.98, H-5) was revealed as equatorial on the basis of its coupling constants (5.2, 5.7, 5.9 Hz), therefore, stereochemistry of quinic acid moiety of **1** was demonstrated as alternative-chair form of authentic 3-CQA (Figure 13). The NMR data of quinic acid region of **1** were in fair agreement with a previous paper

(Wenzl et al., 2000). In their study, stereochemistry of quinic acid molecule of 1,3-di-*O-trans*-feruloylquinic acid isolated from *Brachiaria* species was characterized as alternative-chair form, and proton at the 3-position showed large diaxial coupling constant.

Compound **2** was also revealed to be a conformational isomer of 3-CQA on the basis of its spectral data and acetylation. In the ^{1}H NMR of **2** (Table 1), an oxymethine proton at the 5-position (δ_H 4.07) showed a relatively large vicinal coupling constant (7.4 Hz) coupled with another oxymethine proton (δ_H 3.70, H-4). These protons are assigned to be diaxial. Based on these orientations, quinic acid moiety of **2** was presumed to be boat form, in which case carboxy group at the 1-position and oxymetine proton at the 4-position of which are in axial orientation. To confirm this prediction, nuclear overhauser effect (NOE) experiment on 3-CQA and **2** was carried out. When oxymethine proton at the 4-position was irradiated, the enhancement of two oxymethylene proton (H-2β and H-6β) signals of **2** are apparently smaller than those of 3-CQA (Figure 14). These results supported the presumed stereochemistry of **2**. In the complete boat form, the angles between oxymethine proton at the 5-position and methylene proton (H-6α), as well as oxymethine proton at the 3-position and methylene proton (H-2β), should be 0° corresponding to show large vicinal coupling constant (>8 Hz) (Karplus, 1963). However, these protons showed somewhat smaller coupling constant as 5.1–7.7 Hz than the expected value. In general, boat form of cyclohexane ring is energetically costly than skewed boat form, in addition, a bulky substituent such as caffeoyl group at the 3-position is axial; hence, it was presumed that the cyclohexane ring of quinic acid exists as a skewed boat form having some angle between these coupled protons (Figure 15). Among the natural occurring quinic acid derivatives, stereochemistry of cyclohexane ring of 3-*O*-caffeoyl-muco-quinic acid isolated from *Asimina triloba* was clarified as a skewed boat form (Haribal et al., 1998). In their study, two sets of coupled protons of quinic acid molecule of 3-*O*-caffeoyl-muco-quinic acid, which are diaxial on the same side, also showed somewhat smaller coupling constant values as 5.7–7.5 Hz. On the basis of this evidence, it was concluded that stereochemistry of quinic acid moiety of **2** is skewed boat form (Figure 13).

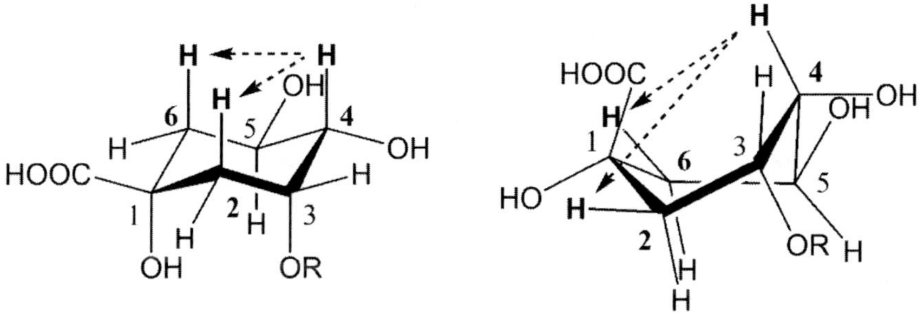

R = caffeoyl

chair form (3-CQA)

skewed boat form (**2**)

Figure 14. Nuclear overhauser effect of 3-CQA and compound **2.**

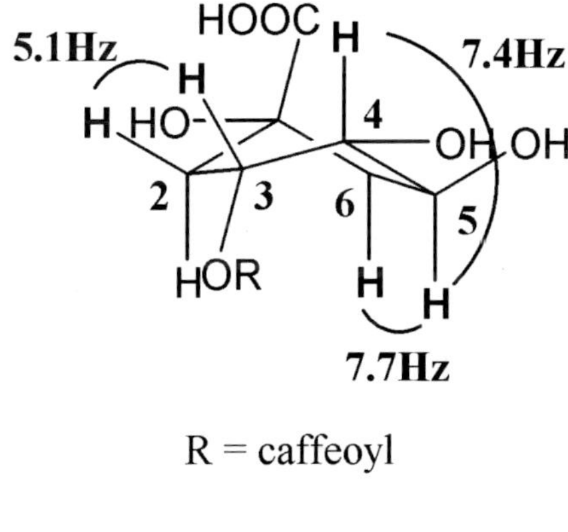

skewed boat form (**2**)

Figure 15. Vicinal coupling of oxymethine and methylene proton of compound **2.**

Novel Compounds

Compound **18** was obtained as a white powder with a negative optical rotation ($[\alpha]^{28}_{D}$ –9.7°). ^{1}H and ^{13}C NMR experiments revealed the presence of a 1,2-disubstituted benzene ring, one methylene group, one quaternary carbon, and two carboxyl groups. The downfield shift of aromatic carbon at the 9-position indicated substitution with an oxygen atom. In an HMBC experiment (Figure 16), long-range correlations were observed between an aromatic proton (H-5) and a quaternary carbon (C-4), methylene protons (H-3) and two carboxyl carbons (C-2 and C-11), and methylene protons (H-3) and an aromatic carbon (C-10). In these results, it was determined that the quaternary carbon (C-4), attached to the aromatic carbon at the 10 position, was substituted with methylene (C-3) and carboxyl carbons (C-11). The IR spectrum of **18** showed absorption at 1736 cm^{-1} (δ-lactone), and the negative HR-FABMS spectrum showed an $[M-H]^-$ peak at *m/z* 206.0482 corresponding to $C_{10}H_9NO_4$. On the basis of these IR and MS analyses, it was established that the carboxyl group at the 2-position is esterified with the hydroxyl group at the 9-position to form a δ-lactone, and an amino group is attached to quaternary carbon (C-4). In addition, this compound was positive in the ninhydrin reaction; hence, the planar structure of **18** was characterized as shown in Figure 16 and this compound was identified as 4-amino-4-carboxychroman-2-one.

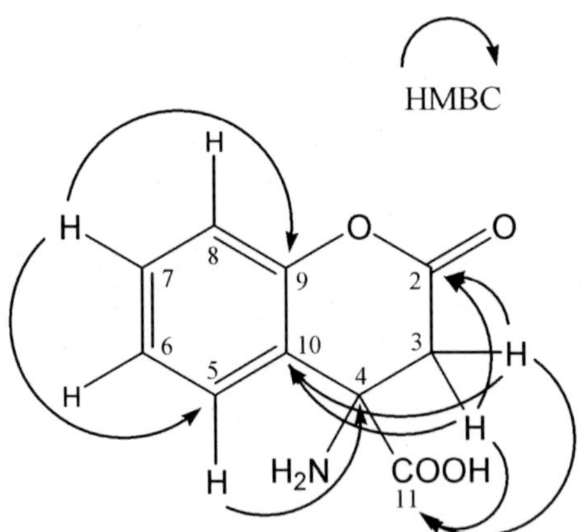

Figure 16. Chemical Structue of compound **18.**

Compound **22** was obtained as a colorless powder with a negative optical rotation ($[\alpha]^{23}_D$ −15.0°), and the negative HR-FABMS spectrum showed an $[M-H]^-$ peak at *m/z* 297.1340 corresponding to $C_{15}H_{22}O_6$. The ^1H and ^{13}C NMR spectra revealed the presence of three tertiary methyls, two methylenes, one of which was expected to bear oxygen atom, two oxymethines, conjugated two double bonds, and one carbonyl and three quaternary carbons. Two olefin protons (H-4 and H-5) showed a trans coupling, and one tertiary methyl (3-CH_3) was coupled with another olefin proton (H-2). On the basis of the HMQC and HMBC experiments, three quaternary carbons were assigned at C-1′, C-5′ and C-8′ together with the assignment of two oxymethines (H-3′ and H-4′), and downfield shift of C-1′ and C-8′ carbon suggested the substitution of oxygen atom. In addition, one methylene group was assigned at the 2′-position, and another oxymethylene protons (H-6′) showed log-range correlations with C-1′ and C-8′ carbons, indicating the presence of cyclohexane ring which has [3,2,1] bicyclo structure as shown in Figure 17. Furthermore, the olefin proton (H-4) also showed a long-range correlation with C-8′, together with a correlation between another olefin proton at the 2-position and carbonyl carbon (C-1). In these results, it was elucidated that the carbon at the 8′-position of bicyclo ring is substituted with diene, which is attached with the carbonyl carbon at the 1-position.

In the NOESY experiment, significant NOEs were observed between H-2 and CH_3 at the 3-position, CH_3 at the 3-position and H-5, and H-5 and H-2′ax, H-4′ax, suggesting the stereochemistry as shown in Figure 17, which bears *trans-cis* conjugated diene at the 8′-position in axial orientation. Therefore, this compound was identified as *rel*-5-(1*R*,5*S*-dimethyl-3*R*,4*R*,8*S*-trihydroxy-7-oxabicyclo[3,2,1]-oct-8-yl)-3-methyl-2*Z*,4*E*-pentadienoic acid (Figure 18).

The ^{1}H and ^{13}C NMR spectra of **23** resembled those of compound **22**. However, they showed one carbonyl carbon signal at the 6-position instead of methylene protons and carbon signals of **22**. In addition, HR-FABMS spectrum showed an $[M-H]^-$ peak at *m/z* 311.1128 corresponding to $C_{15}H_{20}O_7$, therefore, compound **23** was identified as *rel*-5-(1*R*,5*S*-dimethyl-3*R*,4*R*,8*S*-trihydroxy-7-oxa-6-oxobicyclo[3,2,1]oct-8-yl)-3-methyl-2*Z*,4*E*-pentadienoic acid (Figure 18).

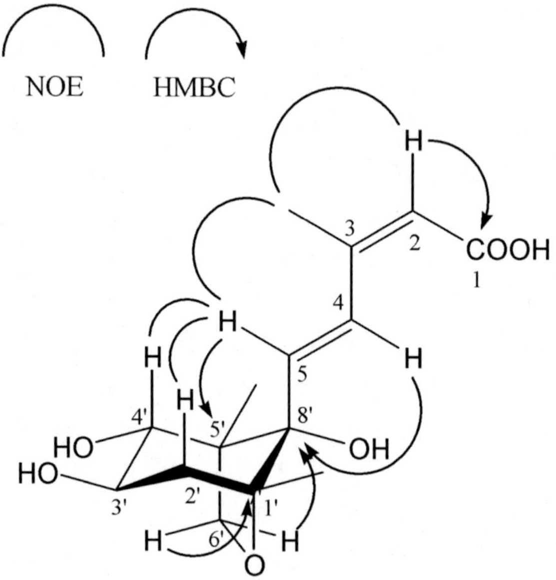

Figure 17. HMBC correlations and NOEs of compound **22.**

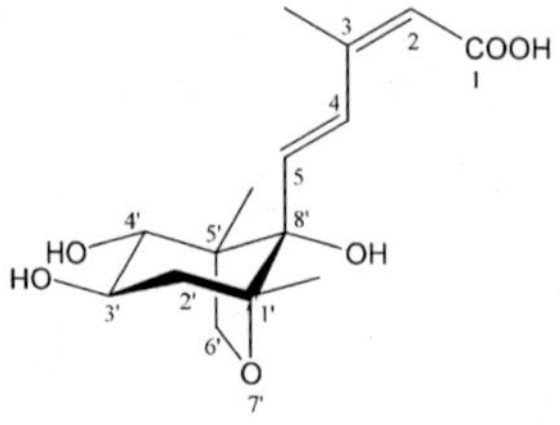

22

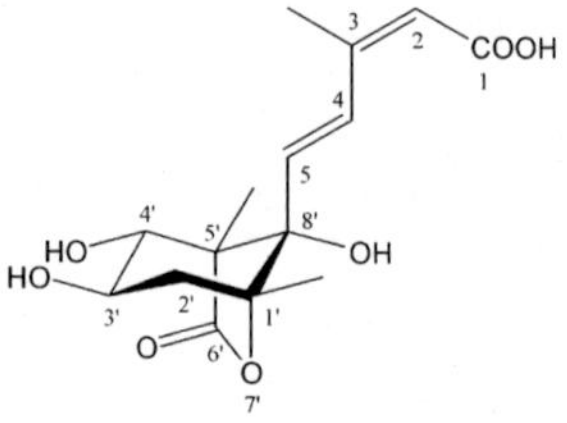

23

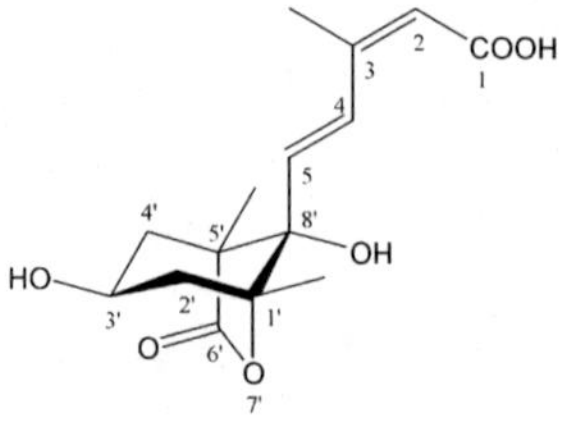

24

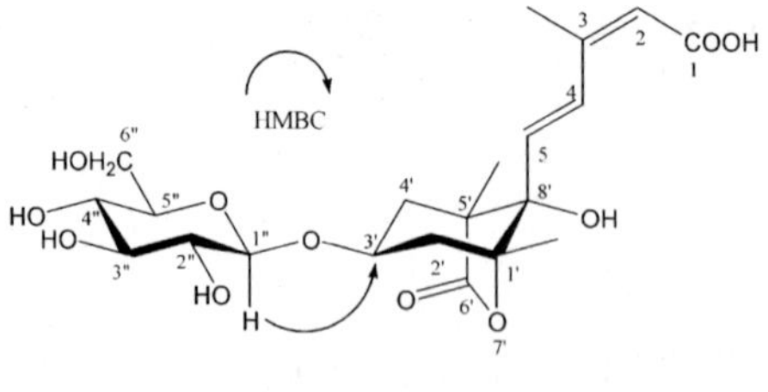

25

Figure 18. Chemical structures of compound **22–25.**

The ^{1}H and ^{13}C NMR spectra of **24** resembled those of compound **23**. However, they showed one methylene proton and carbon signals at the 4′-position instead of oxymethine proton and carbon signals of **23**. In addition,

HR-FABMS spectrum showed an $[M-H]^-$ peak at *m/z* 295.1196 corresponding to $C_{15}H_{20}O_6$, therefore, compound **24** was identified as *rel*-5-(3*S*,8*S*-dihydroxy-1*R*,5*S*-dimethyl-7-oxa-6-oxobicyclo[3,2,1]oct-8-yl)-3-methyl-2*Z*,4*E*-pentadienoic acid (Figure 18).

The ^1H and ^{13}C NMR spectrum of **25** revealed the presence of same structure of compound **24** and β-D-glucose moiety. In the HMBC experiment, long-range correlation was observed between anomeric proton of β-D-glucopyranose (H-1″) and the carbon at the 3′-position of bicyclo ring. In addition, HR-FABMS spectrum showed an $[M-H]^-$ peak at *m/z* 457.1716 corresponding to $C_{21}H_{30}O_{11}$, therefore, compound **25** was identified as *rel*-5-(3*S*,8*S*-dihydroxy-1*R*,5*S*-dimethyl-7-oxa-6-oxobicyclo [3,2,1]oct-8-yl)-3-methyl- 2*Z*,4*E*-pentadienoic acid 3′-*O*-β-D-glucopyranoside (Figure 18).

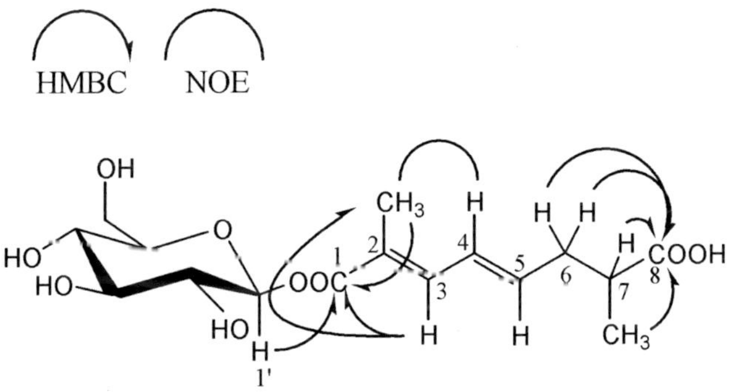

Figure 19. Significant long range correlations in the HMBC spectrum and significant NOEs in the NOESY spectrum of compound **30.**

Compound **30** exhibited an optical rotation of –7.3° and an $[M-H]^-$ peak at *m/z* 359.1327 in good agreement with the molecular formula $C_{16}H_{24}O_9$ by measurement of the negative mode HR-FABMS. The UV spectrum of **30** showed an absorption maximum at 268 nm, suggesting the presence of a conjugated diene group, and the IR spectrum revealed absorption bands at 1700 and 1637 cm^{-1} due to carboxylic functions. The ^1H NMR spectrum of **30** showed two methyl [δ_H 1.16 (*d*, J = 6.8 Hz) and 1.16 (*d*, J = 1.2 Hz)], a methylene [δ_H 2.35 (*ddd*, J = 6.8, 8.5, 16.4 Hz) and

2.53 (*ddd*, J = 6.8, 7.3, 16.4 Hz)], a methine [δ_H 2.54 (*ddd*, J = 6.8, 6.8, 8.5 Hz)], and three olefinic [δ_H 6.12 (*ddd*, J = 6.8, 7.3, 15.0 Hz), 6.50 (*dd*, J = 11.3, 15.0), and 7.28 (*dd*, J = 1.2, 11.3)] protons, and the ^{13}C NMR spectrum indicated two carboxyl (δ_C 168.6 and 179.8) carbons along with four olefinic (δ_C 126.0, 128.9, 141.2, and 142.0), two methyl (δ_C 12.6 and 17.3), a methylene (δ_C 38.2) and a methine (δ 40.7) carbons. ^{1}H-^{1}H correlation spectroscopy (^{1}H-^{1}H COSY) measurement established the sequence of CH-CH=CH-CH_2-CH-CH_3 and an *E*-form of olefin moiety in the molecule. In heteronuclear multiple bond correlation (HMBC) experiment, long-range correlations were observed between the methyl (δ_H 1.16), methylene, and methine protons and the carboxyl carbon (δ_C 179.8), between the methyl (δ_H 1.93) and the olefinic (δ_H 7.28) protons and the carboxyl carbon (δ_C 168.0), and between the olefinic proton (δ_H 7.28) and the methyl carbon (δ_C 12.6), indicating the presence of 2,7-dimethyl-2,4-octadienedioic acid moiety. In addition, ^{1}H and ^{13}C NMR spectra

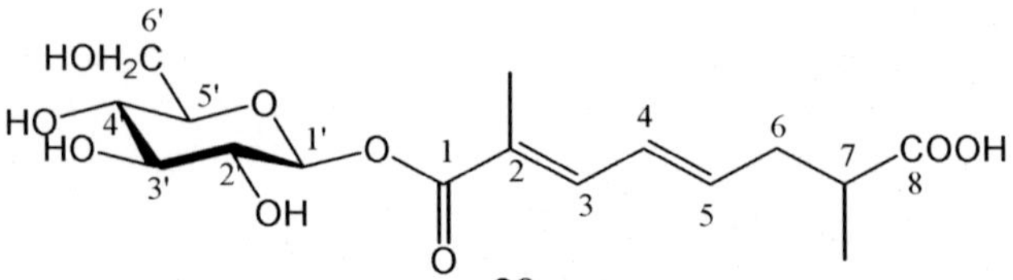

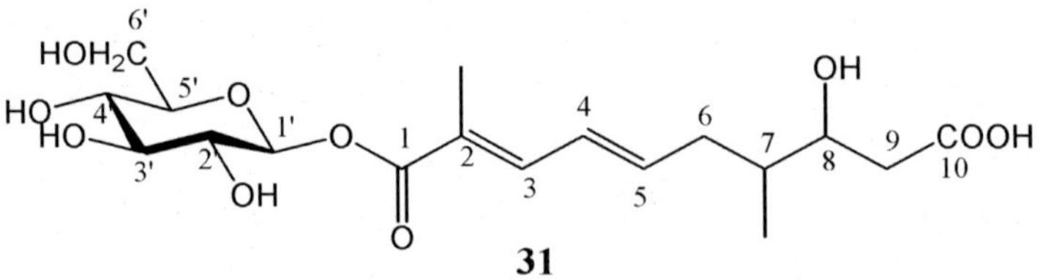

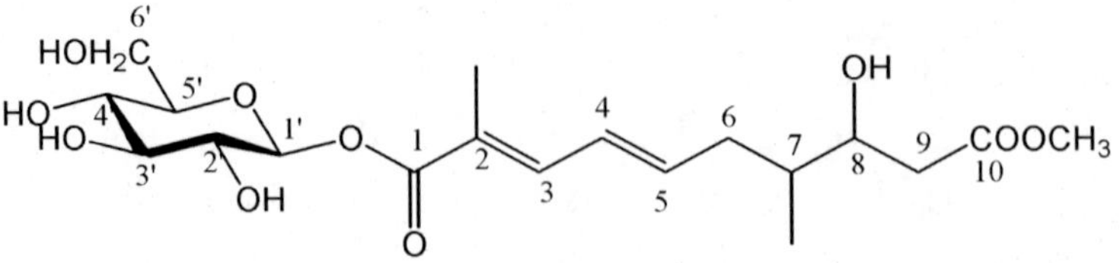

Figure 20. Chemical structures of compound **30–32.**

showed the signals of a β-D-glucopyranosyl part, and HMBC measurement also showed long-range correlation between the anomeric proton [δ_H 5.53 (*d*, J = 7.8 Hz)] and the carboxyl carbon (δ_C 168.6), suggesting the glucosyl esterification at 1-COOH. Furthermore, 2*E*,4*E*-form were confirmed on the basis of the nuclear overhauser and exchange spectroscopy (NOESY) experiment, which showed a correlation between the olefinic proton at 4-position and 2-methyl protons (Figure 19). On the basis of these results, compound **30** was determined to be β-D-glucopyranosyl 7-carboxy-2-methyl-2*E*,4*E*-octadienate (Figure 20), which is a glucosyl ester of 2,7-dimethyl-2*E*,4*E*-octadiene-1,8-dicarboxylic acid (29).

Compound **31** showed an $[M-H]^-$ peak at *m/z* 403.1610 corresponding to the molecular formula $C_{18}H_{28}O_{10}$ in the negative mode HR-FABMS. The ^{1}H and ^{13}C NMR spectra of **31** were resembled to those of **30**, showing additional signals of oxymethine [δ_H 3.77 (*ddd*, J = 3.5, 6.3, 9.0 Hz)] and methylene [δ_H 2.43 (*dd*, J = 3.5, 15.2 Hz) and 2.27 (dd, J = 9.0, 15.2)] protons and carbons (δ_C 41.1 and 73.5) due to C-8 and 9 (Figure 20). Significant correlations were observed between these oxymethine and methylene protons and carboxyl carbon at 10-position in HMBC experiment, and between methine proton [δ_H 1.69 (*m*)], the methylene protons and the oxymethine proton in ^{1}H-^{1}H COSY measurement. In addition, NOESY experiment indicated correlations between methylene protons (H-9α and H-9β) and oxymethine proton (H-8) and between oxymethine proton (H-8) and methine proton (H-7), thus, compound **31** was concluded to be β-D-glucopyranosyl 9-carboxy-8-hydroxy-2,7-dimethyl-2*E*,4*E*-nonadienate. In an earlier study, the aglycon of **31** was isolated from the seed of *Phaseolus multiflorus* (Macmillan and Suter, 1967).

Compound **32** had a molecular formula $C_{19}H_{30}O_{10}$, which was determined by an $[M-H]^-$ peak at *m/z* 417.747 in the negative mode HR-FABMS. ^{1}H and ^{13}C NMR spectra of **32** were very similar to those of **31**, however, additional signals due to a methoxyl function were observed at δ_H 3.68 (*d*, J = 8.0Hz) in the ^{1}H NMR and at δ_C 52.2 in the ^{13}C NMR spectra. Furthermore, the methoxyl protons showed a correlation to the carboxyl carbon at 10-position in HMBC measurement, therefore, compound **32** was determined to be 8-hydroxy-2,7-dimethyl-2*E*,4*E*-decadienedioic acid 1-β-D-glucopyranyl ester 10-methyl ester (Figure 20), which is a methyl ester of **31**. Furthermore, optical rotation and ^{1}H NMR spectrum of artificial methyl ester of **31**, which is obtained by esterification with trimethylsilyldiazomethane, showed completely agreement with those of **32**, indicating that these two compounds have identical stereochemistry. Nevertheless, absolute stereochemistry of these compounds is unknown in this study, because methoxytrifluoromethylphenylacetyl (MTPA) ester at C-8 of **31** in order to apply Mosher's method could not be obtained.

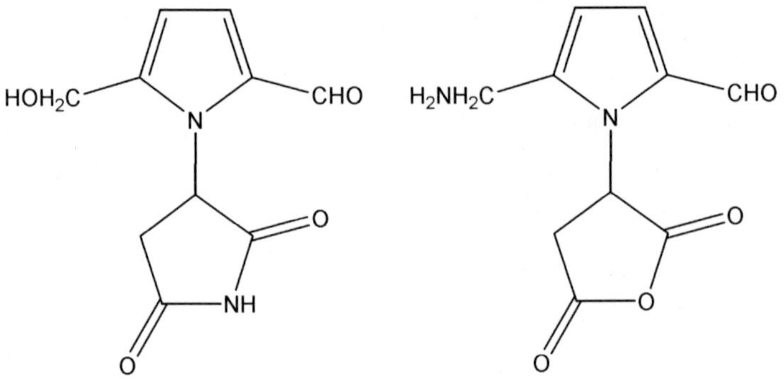

A B

Figure 21. Presumed structure of compound **37.**

Compound **37** was obtained as a white powder with a negative optical rotation ($[\alpha]^{25}_{D}$ −89.5°). ^{1}H and ^{13}C NMR experiments revealed the presence of pyrrole ring, which has hydroxymethyl and aldehyde group, and succinamid moiety, and it was predicted that this compound should be an ester of these molecules. However, in the ^{1}H NMR analysis, acetylation of compound **37** led to show the downfield shift of hydroxymethylene protons (H-7); hence, it was suggested that this hydroxyl group was not esterified. In the HMBC experiment, long-range correlations were observed between H-3′ and C-2, and H-3′ and C-5. Furthermore, negative HR-FABMS spectrum showed an $[M{-}H]^{-}$ peak at *m*/*z* 221.0566 corresponding to $C_{10}H_{10}N_2O_4$; hence, two structures were presumed for compound **37** (Figure 21). However, in the ^{1}H NMR analysis of **37**, two broad proton signals were assigned as hydroxyl proton at the 7-position and amine proton at the 1′-position. Therefore, the planar structure of **37** was characterized as shown in Figure 21A, and this compound was identified as 2-(5-hydroxymethyl-2′,5′-dioxo-2′,3′,4′,5′-tetrahydro-1′*H*-1,3′-bipyrrole)-carbaldehyde.

ANTIOXIDANT ACTIVITY OF ISOLATED COMPOUNDS

Caffeoylquinic Acid Isomers

Oil Stability Index Method

To compare the antioxidant activity of CQA isomers (3-, 4-, and 5-CQA), oil stability index (OSI) method was carried out. As shown in Table 9, the addition of these isomers extended the OSI value compared with that of control (5.50). The values of CQA isomers were close as showing 13.83−15.15, and it seemed that the position of esterification on quinic acid molecule with caffeic acid have no influence on their antioxidant activity.

Table 9. Inhibitory effect against oxidation of methyl linoleate of caffeoylquinic acid isomers and related compounds

	OSI value (hr.)[a]
3-*O*-caffeoylquinic acid (3-CQA)	14.03 ± 1.23
4-*O*-caffeoylquinic acid (4-CQA)	13.83 ± 1.06
5-*O*-caffeoylquinic acid (5-CQA)	15.15 ± 0.39
caffeic acid	46.10 ± 1.50
α-tocopherol	37.83 ± 1.85
BHT	22.85 ± 0.35
control	5.50 ± 0.25

[a] Each sample was measured at 1μmol/5g of oil in triplicate and the data are expressed as the mean values ± the standard deviation. OSI value is defined as the point of maximum change of the rate of oxidation.

On the other hand, the OSI value of caffeic acid was 46.1, which was much higher than those of CQA isomers. In a study of Chen and Ho (1997), antioxidant activity of caffeic acid was higher than 5-CQA when lard was used as a lipid substrate. In contrast, antioxidant activity of caffeic acid and 5-CQA were almost the same on the Trolox equivalent antioxidant activity (TEAC; Rice-Evans et al., 1997) and against oxidation of linoleic acid (Morishita and Kido, 1995). The disparity among these assays might depend on the difference of such as substrate, solubility of sample and measurement temperature.

O_2^- Scavenging Activity and ORAC

The scavenging activity on O_2^- of CQA isomers was also evaluated by direct ESR measurement. As shown in Table 10, the scavenging ratio of 5-CQA at a concentration of 50 μM was 30.1%. This result presented fairly agreement with a previous study, in which 5-CQA showed 53% of scavenging ratio on O_2^- at 100 μM (Tsuchiya et al., 1996).

Table 10. Scavenging activity on O_2^- and oxygen radical absorbance capacity (ORAC) of caffeoylquinic acid isomers and related compounds

	Scavenging ratio of O_2^{-a} (%)	ORAC[b] (unit/μmol)
3-*O*-caffeoylquinic acid (3-CQA)	31.3 ± 3.3	3.89 ± 0.52
4-*O*-caffeoylquinic acid (4-CQA)	37.0 ± 2.2	3.99 ± 0.12
5-O-caffeoylquinic acid (5-CQA)	30.1 ± 5.2	3.39 ± 0.02
caffeic acid	41.1 ± 5.5	2.68 ± 0.05
L-ascorbic acid	47.3 ± 1.1	–[c]

[a] Each sample was measured at 50μM in triplicate and scavenging ratio of O_2^- are expressed as mean values ± the standard deviation. [b] ORAC values are expressed as 1 unit for 1 μmol of Trolox equivalent. [c] not measured.

On the other hand, caffeic acid showed somewhat stronger activity (41.1%) than caffeoylquinic acid isomers. Chen and Ho (1997) compared the free radical scavenging effect of 5-CQA and caffeic acid using the stable 1,1-diphenyl-2-pycrylhydrazyl (DPPH) radicals. In their study, the activity of caffeic acid was one and a half times as high as that of 5-CQA showing similar tendency as this study. In contrast, ORAC of caffeic acid was smaller than those of CQA isomers. This discrepancy might be attributable to the difference of radical species.

When compared three CQA isomers, they exhibited almost the same radical scavenging activity (30.1–37.0%) and ORAC (3.39–3.89 unit/μmol). It seemed that the position of esterification on quinic acid molecule with caffeic acid have no influence on their scavenging activity on O_2^- and ORAC.

Conformational Isomers of 3-CQA

ORAC and Ferric Thiocyanate Method

Antioxidant activity on the basis of ORAC of these compounds was also evaluated. The ORAC value of **1** (2.07 ± 0.06 unit/μmol) was significantly ($p<0.01$) smaller than that of 3-CQA (3.89 ± 0.52), on the other hand, the activity of **2** (4.18 ± 0.12) was almost the same as that of 3-CQA (Table 11).

Table 11. Oxygen radical absorbance capacity (ORAC) of conformational isomers of 3-*O*-caffeoylquinic acid

	ORAC (units/μmol)[a]
3-CQA (chair form)	3.89 ± 0.52
compound **1** (alternative-chair form)	2.07 ± 0.06[b]
compound **2** (skewed boat form)	4.18 ± 0.12

[a] ORAC values are expressed as 1 unit for 1 μmol of Trolox equivalent. Each sample was measured in triplicate and the data are presented as the mean values ± the standard deviations. [b] Significantly smaller than the value of 3-CQA ($p<0.01$).

In the ferric thiocyanate method assay, **1** also showed relatively lower antioxidant activity than **2** and 3-CQA against oxidation of linoleic acid (Figure 22). It seems that these differences of antioxidant activity of these conformers are dependent on their stereochemistry of quinic acid. In the recent study, ORAC of 5-*O*-caffeoylquinic acid was 1.5 times higher than that of caffeic acid alone (Prior and Cao, 1999a). Quinic acid and its stereochemistry might take part in the antioxidant activity of caffeoylquinic acid isomers.

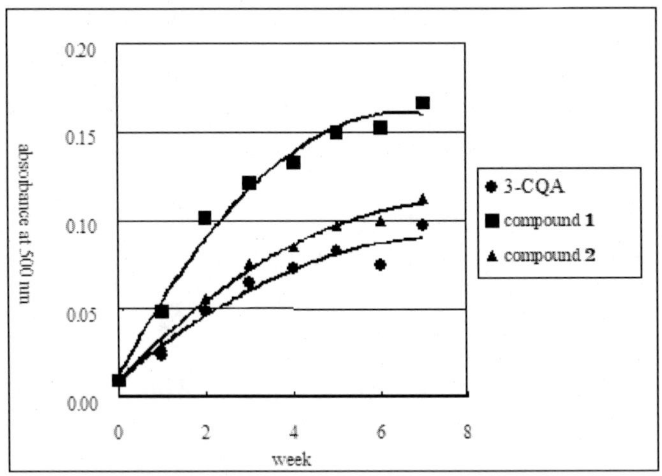

Figure 22. Antioxidant activity of conformational isomers of 3-CQA on the basis of ferric thiocyanate method.

Other Isolated Compounds

ORAC

Antioxidant activity of other isolated compounds were evaluated on the basis of ORAC. Phenolic compounds that have *ortho*-diphenol structure, such as 3- and 4-CQA methyl esters (**4**, **5**), caffeic acid (**7**) and its ester (**8**), protocatechuic acid (**9**), and rutin (**39**) showed high ORAC values as 2.68–5.30 unit/μmol. Mono hydroxyphenyl compounds such as *p*-coumaric acid (**9**), coniferin (**10**), ferulic acid β-D-glucopyranoside (**11**), vanillic acid β-D-glucoside (**17**), 4-amino-4-carboxychoroman-2-one (**18**), magnolioside (**20**), and (+)-pinoresinol *O*-β-D-glucopyranoside (**33**) showed lower ORAC as 0.34–1.54 unit/μmol than ortho-diphenol group. However, vanillic acid (**16**), scopoletin (**19**), and 9-(β-D-glucopyranosyl)-

7-(4-hydroxy-3-methoxyphenyl)-1-(3-hydroxypropyl)-3-methoxy-(7*R*,8*S*)-dihydrobenzofuran (**34**) showed high ORAC as 2.33–4.68 unit/μmol, which are as same as ortho-diphenol compounds (Table 12).

Table 12. Oxygen radical absorbance capacity (ORAC) of hydroxycinnamic acid, benzoic acid, chroman and coumarin, and lignan compounds

	ORAC (unit/μmol)
3-CQA [b] methyl ester (**5**)	3.03 ± 0.00[a]
4-CQA [c] methyl ester (**6**)	3.65 ± 0.07
caffeic acid (**7**)	2.68 ± 0.05
caffeic acid methyl ester (**8**)	3.31 ± 0.03
p-coumaric acid (**9**)	1.04 ± 0.01
coniferin (**10**)	1.54 ± 0.02
ferulic acid β-D-glucopyranoside (**11**)	1.09 ± 0.00
protocatechuic acid (**15**)	3.28 ± 0.03
vanillic acid (**16**)	4.68 ± 0.05
vanillic acid β-D-glucopyranoside (**17**)	0.35 ± 0.08
4-amino-4-carboxychroman-2-one (**18**)	0.32 ± 0.00
scopoletin (**19**)	2.72 ± 0.03
magnolioside (**20**)	0.51 ± 0.02
(+)-pinoresinol *O*-β-D-glucopyranoside (**33**)	1.09 ± 0.03
9-(β-D-glucopyranosyl)-7-(4-hydroxy-3- methoxyphenyl)-1-(3-hydroxypropyl)-3-methoxy- (7*R*,8*S*)-dihydrobenzofuran (**34**)	2.33 ± 0.02
rutin (**39**)	5.30 ± 0.10

[a] ORAC values are expressed as 1 unit for 1 μmol of Trolox equivalent. Each sample was measured in triplicate and data are presented as mean values ± the standard deviations. [b] 3-CQA, 3-O-caffeoylquinic acid. [c] 4-CQA, 4-O-caffeoylquinic acid.

Other non-phenolic compounds, such as abscisic acid related compounds and others, showed low ORAC as <1 unit/μmol (Table 13). It was predicted that phenolic compounds strongly contribute to the antioxidant activity of prunes, and non-phenolic compounds also associated with the activity.

Table 13. Oxygen radical absorbance capacity (ORAC) of abscisic acid related and other compounds

	ORAC (unit/μmol)
r el-5-(1*R*,5*S*-dimethyl-3*R*,4*R*,8*S*-trihydroxy-7- oxabicyclo[3,2,1]-oct-8-yl)-3-methyl-2*Z*,4*E*-pentadienoic acid (**22**)	0.67 ± 0.01[a]
rel-5-(1R,5S-dimethyl-3R,4R,8S-trihydroxy-7- oxa-6-oxobicyclo[3,2,1]oct-8-yl)-3-methyl- 2Z,4E-pentadienoic acid (**23**)	0.34 ± 0.00
rel-5-(3*S*,8*S*-dihydroxy-1*R*,5*S*-dimethyl-7-oxa-6-oxobicyclo[3,2,1]oct-8-yl)-3-methyl-2*Z*,4E- pentadienoic acid (**24**)	0.50 ± 0.01
rel-5-(3*S*,8*S*-dihydroxy-1*R*,5*S*-dimethyl-7-oxa-6-oxobicyclo[3,2,1]oct-8-yl)-3-methyl-2*Z*,4*E*-pentadienoic acid 3′-*O*-β-D-glucopyranoside (**25**)	0.72 ± 0.01
(6*S*,9R)-roseoside (**26**)	0.49 ± 0.00
(+)-abscisic acid (**27**)	0.77 ± 0.02
(+)-β-D-glucopyranosyl abscisate (**28**)	0.41 ± 0.01
hydroxymethylfurfural (**36**)	0.27 ± 0.00
2-(5-hydroxymethyl-2′,5′-dioxo-2′,3′,4′,5′-tetrahydro-1′*H*-1,3′-bipyrrole)-carbaldehyde (**37**)	0.08 ± 0.00
benzyl β-primeveroside (**38**)	0.00 ± 0.00

[a] ORAC values are expressed as 1 unit for 1 μmol of Trolox equivalent. Each sample was measured in triplicate and data are presented as mean values ± the standard deviations.

Synergistic Effect of Compound 18 on ORAC of Caffeoylquinic Acid Isomers

4-Amino-4-carboxychroman-2-one (**18**) showed lower ORAC, 0.32 unit/μmol (Table 12). However, when 1μM of **18** was added to 1μM of 3-*O*-caffeoylquinic acid, the ORAC raised to 5.84 unit/μmol which was higher than an expected value as 4.22 unit/μmol (sum of the data of 3-*O*-caffeoylquinic acid and **18** in Table 14). This compound (**18**) showed a remarkable synergistic effect on each caffeoylquinic acid isomer and the ORAC values were enhanced 1.5–2.9 times (Table 14). It is known that osbekic acid isolated from *Osbeckia chinensis* L. is as an antioxidative synergist to α-tocopherol, methyl gallate, and punicacortein (Su et al., 1987), and ascorbic acid is also well-known as a synergist to α-tocopherol. 4-Amino-4-carboxychroman-2-one (**18**) may contribute to the antioxidant activity of prunes by the synergistic effect on CQA isomers.

Table 14. Synergistic effect of compound 18 on oxygen radical absorbance capacity (ORAC) of caffeoylquinic acid isomers

	ORAC (unit/μmol)
3-*O*-caffeoylquinic acid (3-CQA)	3.89 ± 0.52[a]
4-*O*-caffeoylquinic acid (4-CQA)	3.99 ± 0.12
5-*O*-caffeoylquinic acid (5-CQA)	3.39 ± 0.02
3-CQA + **18** (1 μM + 1 μM)	5.84 ± 0.08
3-CQA + **18** (1 μM + 2 μM)	11.16 ± 2.08
4-CQA + **18** (1 μM + 1 μM)	5.98 ± 0.23
4-CQA + **18** (1 μM + 2 μM)	7.73 ± 0.29
5-CQA + **18** (1 μM + 1 μM)	5.09 ± 0.04
5-CQA + **18** (1 μM + 2 μM)	8.70 ± 0.46

[a] ORAC values are expressed as 1 unit for 1 μmol of Trolox equivalent. Each sample was measured in triplicate and data are presented as mean values ± the standard deviations.

DISCUSSION

Recently, prunes have been recognized as healthy foods; however, the studies concerning the phytochemicals in prunes are small. In this study, the functional substances in prunes such as CQA isomers and organic acids were quantified by improved HPLC analysis. In addition, contribution of CQA isomers to antioxidant activity of prunes were also clarified. Furthermore, isolation, structural elucidation, and evaluation of antioxidant activity of phytochemicals in prunes were also carried out.

In the improved HPLC analysis, it has become apparent that prunes contain relatively high amounts of 4-CQA than other stone fruits and it was characteristic. Concerning the organic acids, it was revealed that prunes contain about 10 times higher contents of quinic acid than the value previously reported. Some investigators suggested that consumption of qunic acid acidify human urine by the means of hippuric acid and prevent bacterial overgrowth of urinary tract infections (Schults, 1984; Kinney and Blount, 1979); therefore, prunes consumption might be useful for controlling urinary tract infections.

It was reported that prunes show high antioxidant activity on the basis of ORAC than other fruits and vegetables (Agricultural Research Service, 1999), and it is also known that the major antioxidant components in prunes are CQA isomers. On this point, it was clarified that the contribution of CQA isomers to the antioxidant activity of prunes was about 30%. In addition, the hydrolyzed residue of prune extract also showed high antioxidant activity, and the existence of conjugated compounds such as proanthocyanidin was suggested.

Isolation, structural elucidation, and evaluation of antioxidant activity of prune components were also carried out. Two conformational isomers of 3-CQA, four novel sesquiterpenoids, three novel monoterpene derivatives, a novel chromanone and a novel bipyrrole were isolated and the structure of them were elucidated on the basis of NMR and MS analyses. Eight hydroxycinnamic acid, three benzoic acid, three abscisic acid related, three lignan, and two coumarin compounds were isolated from prunes for the first time. These isolated phenolic compounds showed strong antioxidant activity, and a novel chromanone exhibited remarkable synergistic effect on antioxidant activity of CQA isomers. It seemed that the antioxidant activity of prunes was highly dependent on CQA isomers with a contribution of other phenolic compounds and the new synergist.

Concerning monoterpene derivatives, 2,7-dimethyl-2*E*,4*E*-octadiene-1,8-dicarboxylic acid (ODA; **29**), which is an aglycon of **30**, was reported as a possible by-product of abscisic acid (ABA; **27**) biosynthesis from oxidative cleavage of C_{40} precursors in tomato (Linforth et al., 1987; Figure 23A). This hypothesis seems to be consistency in the case of *Prunus domestica* L., because it is known that ABA concentration rapidly elevate when the plants are subjected to water stress (Bowman et al., 1984), and ABA and its related compounds were also isolated from prunes in this study. However, a later study showed that when tomato plants were supplied with $^{2}H_2O$ for 6 days, ABA was found to be labeled with deuterium atoms, nevertheless, ODA was unlabeled. Based on these results, it is concluded that ODA cannot be a by-product of the biosynthesis of ABA (Milborrow et al., 1988). In their paper, it is presumed that ODA is formed from cleavage of C_{40} precursors, while, precise details of biosynthesis pathway of ABA is unclear. Thereafter, further studies proposed that ABA synthesis involves oxidative cleavage of 9-*cis*-epoxycarotenoid precursor due to enzymatic reactions (Thompson, A. J., 2000a; 2000b; Taylor, I. B., 2000), on the other hand, biosynthesis of ODA is still unknown in our knowledge. In the present study, C_{12}-acid derivatives (**31**, **32**), which has similar structure to ODA (**29**) were characterized from prunes, and 6*S*,9*R*-roseoside (**26**), an abscisic acid related compound which has C_{13}-skelton, was also isolated from this plant. The structures of them suggest that these compounds might be formed from cleavage of C_{40} precursors such as violaxanthin (Figure 23B), hence, further study must be carried out to reveal the biosynthesis pathway of these isolated compounds from the dried fruits of *Prunus domestica* L., in addition to ABA and ODA.

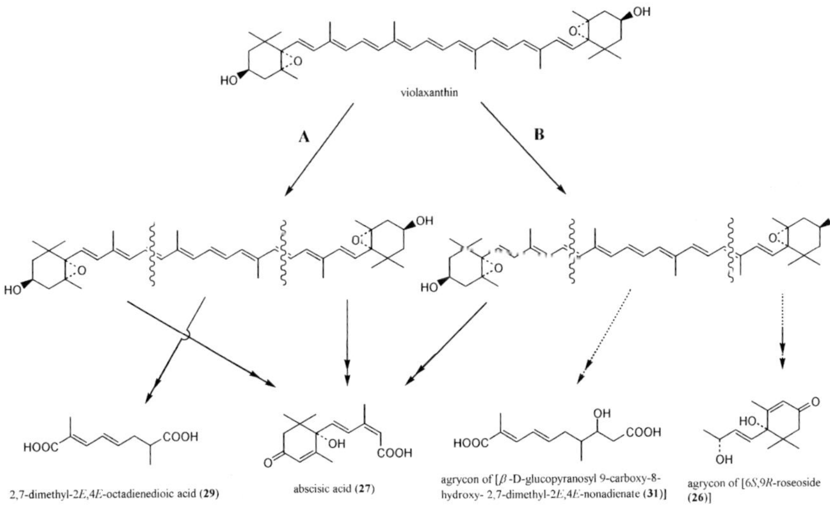

Figure 23. Putative biosynthesis pathway of aglycons of compound **31** and 6*S*,9*R*-roseoside from C_{40} precursors in *Prunus domestica* L.

In conclusion, it was suggested that prunes contain various phytochemicals and are the functional foods for human health on the basis of its high antioxidant activity and large amounts of quinic acid.

REFERENCES

Agricultural Research Service. 1999. "Can food forestall aging?" *Agric. Res.* Feb:14-17.

Ahmed, T., Sadia, H., Batool, S., Janjua, A., and Shuja, F. 2010. "Use of prunes as a control of hypertension." *J. Ayub Med. Coll. Abbottabad.* 22:28–31.

Akoh, C. C. 1994. "Oxidative stability of fat substitutes and vegetable oils by the oxidative stability index method." *J. Am. Oil Chem. Soc.* 71:211-216.

Arjmandi, B. H., Khalil, D. A., Lucas, E. A., Georgis, A., Stoecher, B. J., Hardin, C., Payton, M. E., and Wild, R. A. 2002. "Dried plums improve induces bone formation in postmenopausal women." *J. Womens Health Gend. Based Med.* 11:61-68.

Arjmandi, B. H., Wang, C., Zhang, Y., Lucas, E., Soliman, A., Juma, S., and Stoecker, B. J. 1999. "Prune: its efficacy in prevention of ovarian hormone deficiency-induced bone loss." *J. Bone Mineral Res.* 14:S515.

Bowman, W. R., Linforth, R. S., Rossall, S., and Taylor, I. B. 1984. "Accumulation of an ABA analogue in the wilty tomato mutant, flacca." *Biochem. Genet.* 22:369-78.

California Prune Board. 1997. *California Prune Buyer's Guide*; Pleasanton, CA, p 4.

Cao, G., Booth, L. H., Sadowski, J. A., and Prior, R. L. 1998a. "Increases in human plasma antioxidant capacity after consumption of controlled diets high in fruit and vegetables." *Am. J. Clin. Nutr.* 68:1081-1087.

Cao, G., Russell, R. M., Lischner, N., and Prior, R. L. 1998b. "Serum antioxidant capacity is increased by consumption of strawberries, spinach, red wine or vitamin C in elderly women." *J. Nutr.* 128: 2383-2390.

Cao, G., Verdon, C. P., Wu, A. H. B., Wang, H., and Prior, R. L. 1995. "Automated assay of oxygen radical absorbance capacity with COBAS FARA II." *Clin. Chem.* 41:1738-1744.

Chen, J. H., and Ho, C.-T. 1997. "Antioxidant activities of caffeic acid and its related hydroxycinnamic acid compounds." *J. Agric. Food Chem.* 45:2374-2378.

Chopra, R. N., Nayar, S., C., and Chopra, I. C. 1956. *Glossary of Indian Medical Plants.* C.S.I.R.: New Delhi, India, p 205.

De Moura, J., and Dostal, H. C. 1965. "Nonvolatile acids of prunes." *J. Agric. Food Chem.* 13:33-435.

Deyhim, F., Lucas, E., Brucsewitz, G., Stoecker, B. J., and Arjmandi, H. 1999. "Prune dose-dependently reverses bone loss in ovarian hormone deficient rats." *J. Bone Mineral Res.* 14:S394.

Donovan, J. L., Meyer, A. S., and Waterhouse, A. L. 1998. "Phenolic composition and antioxidant activity of prunes and prune juice (*Prunus domestica*)." *J. Agric. Food Chem.* 46:1247–1252.

Edralin, A. L., Shanil J., Barbara, J. S., and Bahram, H. A. 2000. "Prune suppresses ovariectomy-induced hypercholesterolemia in rats." *J. Nutr. Biochem.* 11:255-259.

Fang, N., Yu, S., and Prior, R. L. 2002. "LC/MS/MS characterization of phenolic constituents in dried plums." *J. Agric. Food Chem.* 50:3579-3585.

Fernandez-Flores, E., Kline, D. A., and Johnson, A. R. 1970. "GLC determination of organic acids in fruits as their trimethylsilyl derivatives." *J. Assoc Off Anal. Chem.* 53:17-20.

Haribal, M., Feeny, P., and Lester, C. C. 1998. "A caffeoylcyclohenane-1-carboxylic acid derivative from *Asimina triloba*." *Phytochem.*, 49:103-108.

Hennig, W., and Herrmann, K. 1980. "Flavonol glycosides of plums of the species *Prunus domestica* L. and *Prunus salicina* Lindley. 12. Phenolics of fruits." *Z. Lebensm. Unters. Forsch.* 171:111-118.

Herrmann, K. 1989. "Occurrence and content of hydroxycinnamic and hydroxybenzoic acid compounds in foods." *Crit. Rev. Food Sci. Nutr.*, 28:315-347.

Hong, V., and Wrolstad, R. E. 1986. "Cranberry juice composition." *J. Assoc. Off. Anal. Chem.* 69:199-207.

Horvat, R. J., Chapman Jr., G. W., Senter, S. D., and Robertson, J. A. 1992. "Comparison of the volatile compounds from several commercial plum cultivars." *J. Sci. Food Agric.* 60:21-23.

Kalt, W., Forney, C. F., Martin, A., and Prior, R. L. 1999. "Antioxidant capacity, vitamin C, phenolics, and anthocyanins after fresh storage of small fruits." *J. Agric. Food Chem.* 47:4638–4644.

Karplus, M. 1963. "Vicinal proton coupling in nuclear magnetic resonance." *J. Am. Chem. Soc.*, 85:2870-2871.

Kikuzaki, H., and Nakatani, N. 1993. "Antioxidant effects of some ginger constituents." *J. Food Sci.* 58:1407-1410.

Kinney, A. B., and Blount, M. 1979. "Effect of cranberry juice on urinary pH." *Nursing Res.* 28: 287-290.

Kono, Y., Kobayashi, K., Tagawa, S. Adachi, K., Ueda, A. Sawa, Y., and Shibata, H. 1997. "Antioxidant activity of polyphenolics in diets. Rate constants of reactions of chlorogenic acid and caffeic acid with reactive species of oxygen and nitrogen." *Biochim. Biophys. Acta.* 1335:335-342.

Linforth, R. S. T., Bowman, W. R., Griffin, D. A., Hedden, P., Marples, B. and Taylor, I. B. 1987. "2,7-Dimethyl-octa-2,4-dienedioic acid, a possible by-product of abscisic acid biosynthesis in the tomato." *Phytochem.* 26:1631-1634.

Macmillan, J., and Suter, P. J. 1967. "The structure of a C_{12}-acid from the seed of *Phaseolus multiflorus*." *Tetrahedron* 23:2417–2419.

Meyer, A. S., Yi, O., Pearson, D. A., Waterhouse, A. L., and Frankel, E. N. 1997. "Inhibition of human low-density lipoprotein oxidation in relation to composition of phenolic antioxidants in grapes (*Vitis unifera*)." *J. Agric. Food Chem.* 45:1638-1643.

Milborrow, B. V., Nonhebel, H. M., and Willows, R. D. 1988. "2,7-Dimethylocta-2,4,-dienedioic acid is not a by-product of abscisic acid biosynthesis." *Plant Sci.* 56:49-53.

Monsefi, M., Parvin, F., and Farzaneh, M. 2013. "Effects of plum extract on skeletal system of fetal and newborn mice." *Med. Princ. Pract.* 221:351-356.

Morishita, H., and Kido, R. 1995. "Antioxidant activities of chlorogenic acids." *Assoc. Sci. Int. Café, Colloq.* 119–124.

Moutounet, M., Dubois, P., and Jouret, C. 1975. "Major volatile compounds in dried plums." *C. R. Acad. Agri.* 61:581-585.

Möller, B., and Herrmann, K. 1983. "Quinic acid esters of hydroxycinnamic acids in stone and pome fruit." *Phytochem.* 22:477-481.

Nagarajan, G. R., and Parmar, V. S. 1977. "Phloracetophenone derivatives in *Prunus domestica*." *Phytochem.* 16:614-615.

Nagels, L., Dongen, W. V., Brucker J. D., and Pooter, H. D. 1980. "High-performance liquid chromatographic separation of naturally occurring esters of phenolic acids." *J. Chromatogr.* 187:181-187.

Nakatani, N., Jitoe, A., Masuda, T., and Yonemori, S. 1991. "Flavonoid constituents of *Zingiber zerumbet* Smith." *Agric. Biol. Chem.* 2:455-460.

Nardini, M., D'Aquino, M., Tomassi, G., Gentili, V., Di Felice, M., and Scaccini, C. 1995. "Inhibition of human low-density lipoprotein oxidation by caffeic acid and other hydroxycinnamic acid derivatives." *Free Radical Biol.* Med., 19:541-552.

Parmar, V. S., Vardhan, A., Nagarajan, G. R., and Jain, R. 1992. "Dihydroflavonols from *Prunus domestica*." *Phytochem.* 31:2185-216.

Pauli, G. F., Kuczkowiak, U., and Nahrstedt, A. 1999. "Solvent effects in the structure dereplication of caffeoylquinic acids." *Magn. Reson. Chem.*, 37:827-836.

Pauli, G. F., Poetsch, F., and Nahrstedt, 1998. "A. Structure assignment of natural quinic acid derivatives using proton nuclear magnetic resonance techniques." *Phytochem. Anal.*, 9:177-185.

Perez, A. G., Olias, R., Espada, J., Olias, J. M., and Sanz, C. 1997. "Rapid determination of sugars, nonvolatile acids, and ascorbic acid in strawberry and other fruits." *J. Agric. Food Chem.* 45:3545-3549.

Pijipers, D., Constant, J. G., and Jansen, K. 1986. "Rosaceae - plum, cherry plum, damson, sloe." In *The Complete Book of Fruit* Multimedia Publications (UK) Ltd.: London, 98-103.

Potterat, O. 1997. "Antioxidants and free radical scavengers of natural origin." *Curr. Org. Chem.*, 1:415-440.

Prior, R. L., and Cao, G. 1999a. "Variability in dietary antioxidant related natural product supplements: the need for methods of standardization." *J. Am. Nutraceut. Assoc.* 2:36–46.

Prior, R. L., and Cao, G. 1999b. "Antioxidant capacity and polyphenolic components of tea: implications for altering *in vivo* antioxidant status." *Proc. Soc. Exp. Biol. Med.* 220:255-261.

Puech, J. L., and Jouret, C. 1974. "Aromatic acids of plum d'Ente and prune d'Agen. (In French.)." *CR. Acad. Agric.* 60:92-95.

Qurechi, I. H., Yaqeenuddin, Yaqeen, Z., Mirza, M., and Qurechi, S. 1988. "Evaluation of the antiemetic action of *Prunus domestic*-Linn." *Pakistan J. Sci. Ind. Res.* 31:774–776.

Raynal, J., Moutounet, M., and Souquet, J.-M. 1989. "Intervention of phenolic compounds in plum technology. 1. Changes during drying." *J. Agric. Food Chem.* 37:1046–1050.

Rice-Evans, C. A., Miller, N. J., and Paganga, G. 1997. "Antioxidant properties of phenolic compounds." *Trends Plant Sci.* 2:152-159.

Rice-Evans, C. A., Miller, N. J., and Paganga, G. 1996. "Structure-antioxidant activity relationships of flavonoids and phenolic acids." *Free Radical Biol. Med.* 20:933-956.

Rodriguez, M. A. R., Oderiz, M. L. V., Hernandez, J. L., Lozano, J. S. 1992. "Determination of vitamin C and organic acid in various fruits by HPLC." *J. Chromat. Sci.* 30:433-437.

Schults, A. 1984. "Efficacy of cranberry juice and ascorbic acid in acidifying the urine in multiple sclerosis subjects." *J. Community Health Nurs.* 1:159-169.

Stacewicz-Sapuntzakis, M., Bowen, P. E., Hussain, E. A., Damayanti-Wood, B. I., and Farnsworth, N. R. 2001. "Chemical composition and potential health effects of prunes: a functional food?" *Crit. Rev. Food Sci. Nutr.* 41:251-286.

Su, J. D., Osawa, T., Kawakishi, S., Namiki, M. 1987. "A novel antioxidative synergist isolated from *Osbeckia chinensis* L." *Agric. Biol. Chem.* 51:3449–3450.

Taylor, I. B., Burbidge, A., and Thompson. A. J. 2000. "Control of abscisic acid synthesis." *J. Exp. Bot.* 51:1563-1574.

Tinker, L. F., Davis, P. A., and Schneeman, B. O. 1994. "Prune fiber or pectin compared with cellulose lowers plasma and liver lipids in rats with diet-induced hyperlipidemia." *J. Nutr.* 124:31-40.

Tinker, L. F., Schneeman, B. O., Davis, P. A., Gallaher, D. D., and Waggoner, C. R. 1991. "Consumption of prunes as a source of dietary fiber in men with mild hypercholesterolemia." *Am. J. Clin. Nutr.* 53:1259-1265.

Thompron, A. J., Jackwon, A. C., Parker R. A., Morpeth, D. R., Burbidge, A., and Taylor, I. B. 2000a. "Abscisic acid biosynthesis in tomato: regulation of zeaxanthin epoxidase and 9-cis-epoxybarotenoid dioxygenase mRNAs by light/dark cycles, water stress and abscisic acid." *Plant Mol. Biol.* 42:833–845.

Thompson, A. J., Jackson, A. C., Symonds, R. C., Mulholland B. J., Dadswell, A. R., Blake, P. S., Burbidge, A., Taylor, I. B. 2000b. "Ectopic expression of a tomato 9-cis-epocycarotenoid dioxygenase gene causes over-production of abscisic acid." *The Plant J.* 23:363–374.

Tsuchiya, T., Suzuki, O., and Igarashi, K. 1996. "Protective effects of chlorogenic acid on paraquat-induced oxidative stress in rats." *Biosci. Biotech. Biochem.* 60:765-768.

Tsuji, M., Harakawa, M., and Komiyama, Y. 1985. "High-performance liquid chromatographic analysis of quinic, malic and shikimic acids in plum fruit using a cation exchange resin. (In Japanese.)" *Nippon Shokuhin Kogyo Gakkaishi* 32:661-663.

Van Gorsel, H., Li, C., Kerbel, E. L., Smits, M., and Kader, A. A. 1992. "Compositional characterization of prune juice." *J. Agric. Food Chem.* 40:784-789.

Vinson, J. A., Su, X., Zubik, L., and Bose, P. 2001. "Phenol antioxidant quantity and quality in foods: fruits." *J. Agric. Food Chem.* 49:5315-5321.

Vinson, J. A., Hao, Y., Su, X., and Zubik, L. 1998. "Phenol antioxidant quantity and quality in foods: vegetables." *J. Agric. Food Chem.* 46:3630-3634.

Wang, S. Y., and Lin, H.-S. 2000. "Antioxidant activity in fruits and leaves of blackberry, raspberry, and strawberry varies with cultivar and developmental stage." *J. Agric. Food Chem.* 48:140–146.

Wang, H., Cao, G., and Prior, R. L. 1996. "Total antioxidant capacity of fruits." *J. Agric. Food Chem.* 44:701–705.

Waterman, P. G., and Mole, S. 1994. "Extraction and chemical quantification." *Analysis of Phenolic Plant Metabolites*, Lawton, J. H., Likens, G. E. Eds; Blackwell Scientific Publications: Cambridge, MA, p 94.

Wenzl, P., Chaves, A. L., Mayer, J. E., Rao, I. M., and Nair, M. G. 2000. "Root of nutrient-deprived *Brachiaria* species accumulate 1,3-di-*O*-trans-feruloylquinic acid." *Phytochem.* 55:389-395.

Williams, R. T. 1971. "The metabolism of certain drugs and food chemicals in man." *Ann. N. Y. Acad. Sci.* 179:141-154.

Wills, R. B. H., Lim, J. S. K., and Greenfield, H. 1986. "Composition of Australian foods. 31. Tropical and sub-tropical fruit." *Food Tech. Australia*. 38:118-123.

Wills, R. B., Scriven, F. M., and Greenfield, H. 1983. "Nutrient composition of stone fruit (*Prunus* spp.) cultivars: apricot, cherry, nectarine, peach and plum." *J. Sci. Food Agric.* 34:1383-1389.

Biographical Sketch

Shin-ichi Kayano

Affiliation:

Department of Nutrition, Faculty of Health Sciences, Kio University

Education:

Osaka City University, Japan. 3/2004 - Graduated.

- PhD. in Science

Osaka City University, Japan. 4/1981–3/1985
- Department of Food Science and Nutrition.

Business Address:

4-2-2 Umami-naka, Koryo-cho, Kitakatsuragi-gun, Nara, 635-0832 Japan

Research and Professional Experience:

-Development of health foods

-Structural elucidation of functional components in plant food materials

Professional Appointments:

- MIKI Corporation, Japan. 4/1985–3/1992
  Member, Research Institute.
- MIKI Corporation, Japan. 4/1992–9/1998
  Researcher, Research Institute.
- MIKI Corporation, Japan. 10/1998–9/2003
  Supervisor, Research Institute.
- Osaka City University, Japan. 7/2003–9/2007
  Visiting Researcher, Graduate School of Human Life Science
- MIKI Corporation, Japan. 10/2003–9/2004
  Supervisor, Development Division.
- Kio University, Japan. 10/2004–3/2007
  Assistant, Faculty of Health Science, Department of Health & Life Science, Health and Nutrition's Course.
  - Engaged as an assistant for experiment practice.
- Kio University, Japan. 4/2007–3/2009
  Associated Professor, Faculty of Health Science, Department of Nutrition
- Kio University, Japan. 4/2007–3/2009
  Associated Professor (M), Graduate School of Health Science

- Osaka City University, Japan. 10/2007–9/2009
  Visiting Associate Professor, Graduate School of Human Life Science.
- Kio University, Japan. 4/2009–Present
  Professor, Faculty of Health Science, Department of Nutrition.
- Kio University, Japan. 4/2009–3/2012
  Professor (M), Graduate School of Health Science.
- Kio University, Japan. 4/2012–Present
  Professor (D), Graduate School of Health Science.

Publications from the Last 3 Years:

1) Antioxidant potential in non-extractable fractions of dried persimmon (*Diospyros kaki* Thunb.). *World Biomedical Frontiers* 2017, http://biomedfrontiers.org/ep-20173-5.
2) Persimmon-derived tannin has bacteriostatic and anti-inflammatory activity in a murine model of Mycobacterium avium complex (MAC) disease. *PLOS ONE* 2017, *12*, e0183489.
3) Antioxidant potential in non-extractable fractions of dried persimmon (*Diospyros kaki* Thunb.). *Journal of Microbial & Biochemical Technology* 2017, *9 Suppl.*, 77.
4) Isoflavone *C*-glycosides isolated from the root of Kudzu (*Pueraria lobata*) and their estrogenic and antimutagenic activities. *Journal of Food and Nutritional Disorders* 2018, *7*, 20.
5) Useful influence on human health of dried persimmon (*Diospyros kaki* Thunb.) (In Japanese). *Shokuhin to Kagaku* 2019, *61*, 14 – 18.

In: *Prunus*
Editor: William Schneider

ISBN: 978-1-53617-755-8

Chapter 2

PRUNUS AVIUM L.: COMPOSITION, ANALYSIS AND HEALTH BENEFITS

Joana Gonçalves[1,2], Ana Y. Simão[1,2,*], Sofia Soares[1,2,*], Carina Gameiro[1,*], Ema Almeida[1,*], Tiago Rosado[1,2,*], Ângelo Luís[1,2,*], PhD, Eugenia Gallardo[1,2], PhD and Ana Paula Duarte[1,2], PhD

[1]Centro de Investigação em Ciências da Saúde,
Universidade da Beira Interior (CICS-UBI), Covilhã, Portugal
[2]Laboratório de Fármaco-Toxicologia, UBIMedical,
Universidade da Beira Interior, Covilhã, Portugal

ABSTRACT

Sweet cherries (*Prunus avium* L.) are amongst the most consumed and appreciated fruits worldwide, and are an excellent source of phytochemicals (melatonin, serotonin, carotenoids and phenolic compounds, including flavonoids and anthocyanins) and nutritive substances (organic acids and sugars). The concentrations of these

* These authors contributed equally to this work.

compounds can vary between different sweet cherry cultivars and in different plant parts thereof.

Recently, this fruit has gained more popularity, as there are many scientific studies that evaluate the effects of sweet cherries as health promoters, emphasizing the health benefits of their bioactive compounds, particularly in what concerns their antioxidant, antimicrobial, antidiabetic, anticancer, anti-inflammatory, anti-neurodegenerative and cardiovascular effects, among others.

This chapter will focus on the description of the composition of these fruits, the main factors that influence their profile of bioactive compounds, the different analytical tools to determine the composition, health benefits associated to their consumption, as well as on the recent findings about their potential therapeutic properties.

Keywords: *Prunus avium* L., nutrients, bioactive compounds, analytical methods, bioactivity, health benefits

Introduction: Chemical Composition and Cultivars

The fruit of *Prunus avium* L., known as sweet cherry, belongs to the *Rosaceae* family, *Prunoideae* subfamily and the *Prunus* genus (Bastos et al. 2015; Ferretti et al. 2010). Sweet cherries are usually consumed raw due to their organoleptic characteristics, but can also be processed as other products such as jellies, liqueurs, juices or jams (Nawirska-Olszańska et al. 2017). This fruit is made up of three parts: exocarp (thin edible outer skin), mesocarp (edible pulp) and endocarp (inedible pit) (Kappel Frank 1996).

Sweet cherry is an important crop, being more prevalent in temperate zones, namely in Europe, North Africa, Asia, Australia and New Zealand and America (Bastos et al. 2015; Commisso et al. 2017). In the past, the USA were the biggest exporter of this fruit, but in recent years, Turkey has become the largest producer of sweet cherry, followed by the USA and then Iran (McCune et al. 2011). There are several cultivars of this fruit, differing in the harvest period, pest resistance, properties such as skin and flesh colour, size, consistency, acidity, sweetness and firmness (Ferretti et al. 2010; Commisso et al. 2017).

These characteristics are due to their biochemical constituents, and therefore this fruit assumes great importance to the scientific community due to its nutritional and bioactive properties (Bastos et al. 2015). Sweet cherries with different characteristics regarding skin colour and consistency, flesh and juice have been described, namely dark red flesh and juice, yellow flesh and clear skin and juice, and light juice and yellow skin (Mulabagal et al. 2009).

The chemical composition of sweet cherry depends on several factors, including cultivar, ripening stage, agricultural practices and environmental conditions (Serradilla et al. 2015). Sweet cherries are mainly composed of water, but they are also an excellent source of nutritive compounds, especially carbohydrates (sugars and fiber), fatty and organic acids, amino acids, vitamins, minerals and phytochemicals such as melatonin, serotonin, carotenoids, phenolic acids (hydroxycinnamic derivatives) and flavonoids (anthocyanins, flavonols and flavan-3-ols) (Hanbali et al. 2013; Usenik, Fabčič, and Štampar 2008; Bastos et al. 2015).

Sweet cherries are mostly composed of water (about 80%), with an energy value of approximately 58 kcal/100 g of fresh fruit and a pH between 3.81 and 3.96 (Vavoura et al. 2015; Bastos et al. 2015). Carbohydrates are among the most common constituents, accounting for about 13.3%; however, when compared to other fruits, sweet cherries have low sugar contents (Belitz, Grosch, and Schieberle 2004). Glucose and fructose are the most abundant monosaccharides, with amounts of 6.59 g and 5.37 g/100 g of fresh fruit, respectively (Papp et al. 2010). Other carbohydrates are also present, namely maltose, sucrose and galactose. These compounds are present in smaller quantities, with their approximate weight being 0.12 g/100 g fresh fruit, 0.15 g/100 g fresh fruit and 0.59 g/100 g fresh fruit, respectively (United States Department of Agriculture 2019). Fibersare also present in sweet cherries, corresponding to about 2.1 g/100 g fresh fruit, pectin being the most abundant (0.4 g/100 g fresh fruit) (Dembitsky et al. 2011). Other macronutrients are amino acids and proteins (Table 1). The amounts of these compounds are about 1.06 g/100 g fresh fruit, and so far isoleucine, threonine, lysine, leucine, cystine, methionine, tyrosine, phenylalanine, aspartic acid, valine, glycine, glutamic acid and serine have been detected (United States Department of Agriculture 2019). In addition, sweet cherry has a low saturated fat content and presents no cholesterol (Bastos et al. 2015). So far, 19 fatty acids have been detected in the fruit, the most abundant being linoleic acid representing about 25% of total fatty acids. Oleic acid is the second most abundant fatty acid, accounting for 23.95% of total fatty acids, while palmitic and α-linolenic acids are present in smaller amounts, around 22.27% and 15.39%, respectively (Bastos et al. 2015). Organic acids such as fumaric, ascorbic, malic, citric, succinic and shikimic are also part of the composition of sweet cherries (Usenik, Fabčič, and Štampar 2008; Demir Taki 2013; Serradilla et al. 2011). These compounds represent 1% of the constituents of the edible fraction of the fruit, the most abundant being malic acid (about 98% of total organic acids), followed by the citric, shikimic and fumaric acids (Usenik, Fabčič, and Štampar 2008).

Sweet cherries are also an excellent source of micronutrients. Minerals are present with a quantity of about 267.56 mg/100 g fresh fruit (United States Department of Agriculture 2019). The main minerals present are potassium, phosphorus, calcium and magnesium, with potassium being the most abundant (United States Department of Agriculture 2019). Iron, cobalt, copper, zinc, manganese, iodine and molybdenum are also present in smaller amounts (United States Department of Agriculture 2019). Vitamins are also present, being vitamin C the most abundant and varying between 7 and 37 mg/100 g fresh fruit. Other vitamins such as B, A, E and K are present as well (United States Department of Agriculture 2019; Lim 2012; Ferretti et al. 2010; Commisso et al. 2017).

Sweet cherries present in their composition a wide array of phytochemicals, namely carotenoids. To date, lutein, β-carotene, α-carotene, β-cryptoxanthin, zeaxanthin and β-apo-8-carotenal have been detected in different sweet cherry cultivars (Demir Taki 2013; Dias, Camões, and Oliveira 2009). Volatile compounds are also present, being responsible for the aroma. These substances are present in small quantities, but so far over 60 volatiles have been identified, for instance (E)-2-hexen-1-ol, benzaldehyde, hexanal and (E)-2-hexenal (Wen et al. 2014). Other important compounds present in sweet cherries are serotonin and melatonin. The serotonin contents range from 2.8 to 37.6 ng/100 g of fresh fruit (González-Gómez et al. 2009), while melatonin is found in smaller amounts (between 1 and 22 ng/100 g fresh fruit), varying from cultivar to cultivar. However, this important phytochemical is present in all parts of the fruit (skin, pulp and stone) (González-Gómez et al. 2009). All these compounds give sweet cherries a high nutritional value, but this fruit is also a source of bioactive molecules, namely phenolic compounds (Figure 1), including phenolic acids (hydroxycinnamic derivatives), flavonoids and anthocyanins, which have several beneficial biological effects (Bastos et al. 2015; M. Wang et al. 2017; Mikulic-Petkovsek et al. 2016). More than 10,000 polyphenols have been identified, the most predominant group being flavonoids (Fürstenberg-Hägg, Zagrobelny, and Bak 2013; D. Del Rio et al. 2010). Within flavonoids, anthocyanins stand out as the most relevant subgroup. These photosensitive compounds are responsible for the characteristic red colour of sweet cherries, and are easily degraded at elevated temperatures (Vermerris and Nicholson 2006). Also, sensory properties such as aroma, astringency and bitter are due to these phenols, which are found in larger amounts on the skin of the fruit (Ferretti et al. 2010; M. Wang et al. 2017; Serrano et al. 2005). The quality of the fruit and its nutritional value is also related to the phenolic composition (Serrano et al. 2005). The most frequently polyphenols in sweet cherries are chlorogenic acid, neochlorogenic acid, epicatechin, rutin, catechin, quercetin-3-*O*-rutinoside, *p*-coumaroylquinic acid, cyanidin-3-glycoside, cyanidine-3-rutinoside, pelargonidine-3-glycoside, peonidine-3-rutinoside and peonidine-3-glycoside (Bastos et al. 2015; Grigoras et al. 2012; Kelebek and Selli 2011; Ieri, Pinelli, and Romani 2012). In addition to these compounds, Bastos et al. (Bastos et al. 2015) have also identified cis-3-*p*-cumaroylquinic acid (derived from phenolic acids), three anthocyanins and six other flavonoids (quercetin-*O*-deoxyhexosylhexoside-*O*-hexoside, naringenin-*O*-hexoside, sakuranine, taxifoline, aromadendrin and taxifoline-*O*-deoxyhexosylhexoside). Thus, the different phenolic compounds and their levels vary in the different sweet cherry cultivars, being the pre- and post-harvest factors also decisive for the amount of phenolic compounds present (M. Wang et al. 2017). The ripening process, which occurs when the colour changes from green to red, is related to the variation of phenolic compounds. At the early stages, when the fruit is still immature, the total phenolic and chlorophyll contents are higher. Up to half of the ripening process, polyphenols decrease and then increase at a later stage. Chlorophyll decreases progressively, while the amount of anthocyanins increases (Ferretti et al. 2010; Serrano et al. 2005; Redondo et al. 2017). Serrano et al. (Serrano et al. 2005) conducted a study in which it was found that the weight and fruit size of *Prunus avium* L. increase up to a certain stage of the ripening process and then remain constant until its final stages. In the same study, it was also possible to verify that during the process, the hardness of the pulp decreases as well as the force necessary to remove the stem (Serrano et al. 2005).

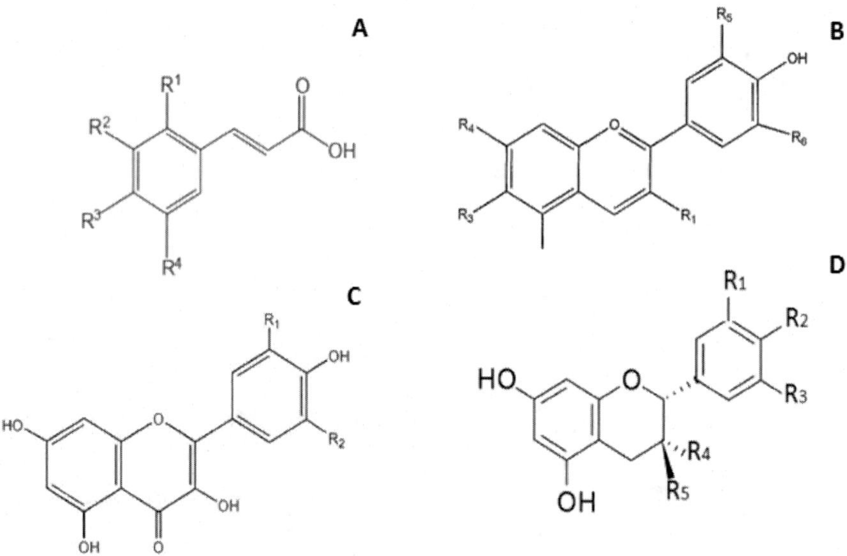

Figure 1. Chemical structure of phenolic acids (A); anthocyanins (B); flavonols (C) and flavan-3-ols (D).

Table 1. Main nutritive compounds of sweet cherries. For more information, refer to USDA FoodData Central (available at https://fdc.nal.usda.gov/fdc-app.html#/?query=sweet%20cherries)

Compounds	Composition*/100 g of sweet cherry	References
Water	82.25 mg	(United States Department of Agriculture 2019)
Energy	57.65 ± 6.85 g	(Bastos et al. 2015)
Total protein	0.42 ± 0.01 g	
Total lipids	0.04 g	
Carbohydrates	13.90 ± 1.72 g	
Fiber	2.1 g*	(United States Department of Agriculture 2019)
Melatonin	0.6-22.4 ng	(González-Gómez et al. 2009)
Serotonin	2.8-37.6 ng	
Glucose	6.02 ± 0.10 g	(Bastos et al. 2015)
Fructose	5.47 ± 0.34 g	
Sorbitol	1.62 ± 0.06 g	
Galactose	0.59 g*	(United States Department of Agriculture 2019)
Sucrose	0.15 g*	
Maltose	0.12 g*	
Tryptophan	3.65-8.27 mg	(Cubero et al. 2010)
Fatty Acids		
Linoleic Acid	25.08 ± 0.07 g	(Bastos et al. 2015)
Oleic Acid	23.95 ± 0.54 g	
Palmitic Acid	22.27 ± 0.77 g	
α-Linolenic acid	15.39 ± 0.20 g	
Malic acid	715.78 ± 0.68 mg	(Bastos et al. 2015)
Citric acid	6.53 ± 0.30 mg	
Fumaric acid	0.37 ± 0.02 mg	
Shikimic acid	0.66-2.67 mg	(Usenik, Fabčič, and Štampar 2008)
Potassium	200 mg	(Ferretti et al. 2010)
Phosphorus	20 mg	
Calcium	14 mg	
Magnesium	10 mg	
Vitamin C	7.0 mg	(Ferretti et al. 2010)
Pantothenic acid	0.2 mg	
Niacin	0.2 mg	
Vitamin A	64 IU	
Vitamin E	0.1 mg	
Vitamin K	2.1 μg	
Lutein + Zeaxanthin	85 μg	
β-Carotene	38 μg	
Choline	6.1 mg	(United States Department of Agriculture 2019)

Legend: * Results expressed as mean values.

However, phenolic compounds levels can be affected by various factors associated with climatic and agronomic conditions called pre-harvest factors (M. Wang et al. 2017; Commisso et al. 2017). Environment temperature, light intensity and type of cultivar are crucial factors that dictate the amount and stability of polyphenols, also affecting the nutritional value of sweet cherries (Ferretti et al. 2010; McCune et al. 2011). Studies have shown that in environments with higher temperatures (25-30°C) and higher sun light exposure, the concentration of anthocyanins increase significantly (Ferretti et al. 2010; McCune et al. 2011). Soil composition, type of fertilization, the water and nutrient supply needed by the plant are also factors that interfere with nutritional composition (Ferretti et al. 2010). Studies conducted on different cultivars of sweet cherries have shown that the phenolic content is different from cultivar to cultivar (Serra et al. 2011). Other conditions that affect polyphenols contents in sweet cherry are post-harvest factors, such as storage and transportation conditions (McCune et al. 2011). It has been found that the total phenolic content of this fruit increases in the days following its purchase, as the ripening process is progressing, decreasing its acidity and increasing the colour intensity due to anthocyanin increase (Ferretti et al. 2010). Studies carried out on different cultivars of sweet cherries also showed that following post-harvest irradiation with UV-B lamp, the anthocyanin content is significantly higher than in fruits irradiated with an UV-A lamp (Arakawa 1993). Gonçalves et al. (B. Gonçalves et al. 2004) conducted a study where it was found that the levels of total phenolic compounds and anthocyanins varied according to the storage conditions. Concerning the cultivars, there are hundreds of commercial sweet cherry cultivars (Usenik, Fabčič, and Štampar 2008; Acero et al. 2019; Chockchaisawasdee et al. 2016; J. Gonçalves, Ramos, Rosado, et al. 2019; Habib et al. 2017), but there are usually only a number of predominant cultivars in a determined country or region; this is what happens for instance with the *Saco* and *Morangão* cultivars, which are traditionally Portuguese (Serradilla et al. 2015; Chockchaisawasdee et al. 2016; J. Gonçalves, Ramos, Rosado, et al. 2019).

Table 2. Sweet cherry cultivars composition (mg/100g fresh weight)

Sweet cherry	Hydroxycinnamic acids*			Anthocyanins*					Flavonols*	Favan-3-ols*		References
cultivars	NcAC	pCqAC	CAc	C3g	C3r	Pl3r	Pn3g	Pn3r	Rut	Cat	Ep	
Burlat	6.8-9.18*	6.4-11.28*	1.1-1.83*	2.3-34.84*	8.28-46.92*	0.2-0.50*	-	0.1-2.31*	4.5	-	3.1	(Serradilla et al. 2015; Chockchaisawasdee et al. 2016)
Sunburst	16.92	9.25	1.62	1.46	21.84	0.41	-	0.94	-	-	-	(Serradilla et al. 2015)
Sweetheart	9.88-41.3*	3.50-64.3*	1.46-2.5*	1.39-1.7*	17.9-22.42*	0.1-0.44*	0.4	0.4-3.83*	2.2-7.56*	-	4.43-6.2*	(J. Gonçalves, Ramos, Rosado, et al. 2019; Serradilla et al. 2015; Chockchaisawasdee et al. 2016)
Lapins	8.7	0,77	1.69	0.10	3.07	0.31	-	0.04	2.1	-	0.43	(Serradilla et al. 2015; Chockchaisawasdee et al. 2016)
Summit	29.0-35.2*	2.49-98.7*	3.13-3.5*	2.5-3.54*	21.6	0.2	0.3	0.3	1.4-4.13*	-	2.65-8.1*	(L. Gao and Mazza 1995; Chockchaisawasdee et al. 2016; J. Gonçalves, Ramos, Rosado, et al. 2019)
Van	17.31-87.5*	4.15-23.0*	3.1-5.77*	1.51-12.0*	20.2-130.2*	0.1-0.7*	0.2-0.63*	0.6-7.4*	1.9-3.55*	4.51	5.3-6.33*	(Chockchaisawasdee et al. 2016; Serradilla et al. 2015; L. Gao and Mazza 1995)
Early Van Compact	11.9	1.2	2.3	0.2	8.06	0.8	-	0.2	3.5	-	1.3	(Chockchaisawasdee et al. 2016)

Legend: NcAC: neochlorogenic acid; pCqAC: p-Coumaroylquinic acid; CAc: Chlorogenic acid; C3g: Cyanidin-3-glucoside; C3r: Cyanidin-3-rutinoside; Pl3r: Pelargonidin-3-O-rutinoside; Pn3g: Peonidin-3-O-glucoside; Pn3r: Peonidin-3-O-rutinoside; Cat: Catechin; Ep: Epicatechin; Rut: Rutin. *Results expressed as mean values.

Among all cultivars of sweet cherries, some are more common than others, such as *Burlat*, *Summit*, *Van*, *Early Van Compact*, *Lapins*, among others. Phenolic compounds present in some cultivars, as well as their reported concentrations, are summarized in Table 2.

BIOACCESSIBILITY AND BIOAVAILABILITY

It has long been reported that phenolic compounds, such as anthocyanins, are not absorbed following oral consumption (Fazzari et al. 2008). It was accepted that polyphenols were degraded in the gastrointestinal tract by bacterial enzymes, and that additionally the glycols could be partially absorbed and biotransformed by bacteria (Fazzari et al. 2008; Bokkenheuser, Shackleton, and Winter 1987). Evidence demonstrated that anthocyanins, as well as their glycosylated forms were absorbed, and were detected in human plasma samples by high performance liquid chromatographic analysis coupled to a photodiode array detector (HPLC-DAD) (Paganga and Rice-Evans 1997). Further studies have shown that glycosylated forms of anthocyanins are absorbed intact (Fazzari et al. 2008; Mülleder, Murkovic, and Pfannhauser 2002).

Thus, several works began to be developed, in order to understand the bioavailability and bioaccessibility of phenolic compounds present in food matrices. So far, only two studies have been developed to assess bioaccessibility and bioavailability in sweet cherries.

Fazzari et al. (Fazzari et al. 2008) evaluated the bioavailability of phenolic compounds present in five cultivars of frozen sweet cherries (*Bing*, *Lapins*, *Skeena*, *Staccato* and *Sweetheart*). Among these five cultivars, two of them (*Bing* and *Lapins*) were in three different maturation states (mature, immature and overmatured). For this purpose, the samples were prepared by adding distilled water to the frozen sweet cherries (sliced and pitted), then crushed, homogenized and acidified with HCl 5M (pH=2) (Fazzari et al. 2008). The extracts were subjected to an *in vitro* digestion process which consisted of an initial digestion with pepsin-HCl under stirring at 37°C to simulate gastric digestion. Pancreatin and bile salts were then added to simulate intestinal digestion. After these reactions the pH was adjusted to 7.7 with 1M $NaHCO_3$ (Fazzari et al. 2008). The authors used tubes with a dialysis membrane to simulate the intestinal wall, where one side of the membrane corresponded to "IN" (accessible to serum) and the other to "OUT" (remains in the gastrointestinal tract) (Fazzari et al. 2008). The digested solutions and the control were incubated at 37°C under shaking for 2h along with the dialysis tubes (Fazzari et al. 2008). After incubation, the material on the "IN" side of the membrane was weighed and the pH measured. After the digestion reaction the amount of total phenols was measured using the Folin-Ciocalteu method (Singleton and Rossi 1965) using gallic acid as standard. The total anthocyanin concentration was also determined using the modified Glories method (Mazza et al. 1999) using cyanidin-3-glucoside as standard. All samples of the digested fractions contained on the "IN" and "OUT" sides were further analysed by HPLC-DAD (Fazzari et al. 2008). The results showed that after digestion the amount of total phenolics had increased when compared to the initial extract. On the other hand, the amount of anthocyanins was not altered during the simulated digestion process (Fazzari et al. 2008). The *Lapins*, *Sweetheart* and *Skeena* cultivars were the varieties with the highest amounts of anthocyanins and polyphenols. The amount of total phenols on the "IN" side ranged from 26% to 30%, while the amount of the same compounds on the "OUT" side ranged from 77% to 100% (Fazzari et al. 2008). Regarding the amount of anthocyanins, the percentages were lower. On the "IN" side of the dialysis membrane the percentage of anthocyanins ranged from 15% to 21%, while in the "OUT" side that percentage ranged from 52% to 67% (Fazzari et al. 2008). In this study, the bioavailability of the polyphenols present in the *Lapins* and *Bing* cultivars was evaluated in the immature, mature and overmatured states, and the cultivars in the immature state had the lowest concentrations of phenolic compounds. The results obtained by the spectrophotometric methods were corroborated by the HPLC analysis of the *Lapins* cultivar in the three maturation states (Fazzari et al. 2008).

A more recent study by Gonçalves et al. (J. Gonçalves, Ramos, Luís, et al. 2019) evaluated the bioaccessibility and bioavailability of the phenolic compounds present in the sweet cherry *Saco* cultivar. For that, the samples of sweet cherry extracts were prepared by lyophilization of the fresh fruit, followed by an extraction process directed to anthocyanins with HCl acidified methanol (0.1%) (J. Gonçalves, Ramos, Luís, et al. 2019). Subsequently, the extract was centrifuged, and the solvent evaporated under reduced pressure. The extracts were subjected to an *in vitro* digestion process previously described by Minekus et al. (Minekus et al. 2014) and Versantvoort et al. (Versantvoort et al. 2005), preparing four digestion solutions (salivary, gastric, duodenal and biliary). One gram of extract was dissolved in deionized water, salivary fluid was added and incubated for 5 min at 37°C while stirring. The gastric fluid was then added and incubated at 37°C while stirring for 2h. Finally, duodenal and bile fluids were added together with $NaHCO_3$ solution (1M) and incubated again at 37°C under stirring for 2h (J. Gonçalves, Ramos, Luís, et al. 2019). At the end of each step, aliquots were collected to quantify the phenolic compounds released from the food matrix and to evaluate their antioxidant activity. The extracts obtained before and after the *in vitro* digestion process were incubated with a monolayer of the cell line derived from a human colon carcinoma (Caco-2) at different times (0.5h, 1h, 2h, 4h and 6h) to evaluate the amount of polyphenols that became bioavailable (J. Gonçalves, Ramos, Luís, et al. 2019). The integrity and permeability of the Caco-2 cell monolayer was evaluated by the transepithelial electrical resistance assay (TEER) and lucifer yellow permeability assay, respectively. Phenolic compounds present in all aliquots collected during the *in vitro* digestion and cell incubation process were quantified using HPLC-DAD equipment (J. Gonçalves, Ramos, Luís, et al. 2019). The antioxidant activity of these same aliquots was determined by DPPH free radical scavenging assay. The results showed that the phenolic compounds of *Prunus avium* L. fruit, after undergoing a simulated digestion process, were absorbed by the cell barrier, becoming bioavailable (J. Gonçalves, Ramos, Luís, et al. 2019). However, these compounds were in lower concentrations than the initial ones. In contrast, when the same undigested sweet cherry extract was brought into contact with the same cell layer only quercetin-3,4'-di-*O*-glycoside become bioavailable (J. Gonçalves, Ramos, Luís, et al. 2019).

Also, the antioxidant activity was higher before the cell monolayer passage, being the percentage of DPPH inhibition of 0% after this same process. In contrast to incubation with the *in vitro* digested extract, the integrity of the cell monolayer was altered, and its permeability increased upon incubation with the crude extract (J. Gonçalves, Ramos, Luís, et al. 2019). This suggests that the compounds present in the initial extract, which interfere with the intercellular junctions, undergo modifications after the digestive process, becoming more bioaccessible, since they are degraded, resulting in smaller compounds. It was possible to verify that the absorption was facilitated, increasing its bioavailability, and these factors are relevant to the cellular integrity and permeability (J. Gonçalves, Ramos, Luís, et al. 2019).

Biological Properties

In the last decade, the scientific interest about this fruit has gained more popularity and there are many studies that evaluate the effects of sweet cherries as health promoters, giving emphasis to the health benefits of their bioactive compounds, particularly in what concerns to their antioxidant, antimicrobial, antidiabetic, anticancer, anti-inflammatory, neuroprotective, and cardiovascular effects, among others. In the next lines we will present the main findings about the health promoting properties of *Prunus avium* L.

Antioxidant

Regarding the antioxidant properties, several studies have been performed on sweet cherries, which demonstrated and allowed to state that, in fact, they have components with antioxidant activity, due to the presence of phenolic compounds, ascorbic acid, β-carotene and vitamin E (Wenzel et al. 2005). In this section, we will emphasize the studies about the antioxidant activity of this fruit.

Usenik et al. (Usenik, Fabčič, and Štampar 2008), investigated the total antioxidant activity of different sweet cherries cultivars at different times of their ripening state using the DPPH method. The main conclusions of this study suggest that there is a correlation between antioxidant activity with both total phenolic content and the content of anthocyanins was cultivar dependent, due to different chemical features of the cultivars.

A study was performed assessing the effects of the different cultivars (*Burlat*, *Van*, *Tragana* and *Mpakirtzeika*), the orchard elevation (low 39-59 m, medium 216 m and high 490-546 m), as well as the effect of the storage (2 or 4 days at 2°C) on the quality and characteristics of sweet cherries, namely the total antioxidant capacity (through DPPH assay). With this research, Faniadis et al. (Faniadis, Drogoudi, and Vasilakakis 2010) showed that the *Tragana* and *Burlat* cultivars had higher antioxidant capacity than the other cultivars. Moreover, they were able to correlate the geographic orchard elevation with the total antioxidant capacity, that is, the higher the elevation, the higher the antioxidant content of the sweet cherries, when comparing to the lower elevation orchards, however, this was not true for the *Van* cultivar. These authors were also able to state that the antioxidant activity is significantly associated with the total phenolic content in the *Tragana*, *Burlat* and *Mpakirtzeika* cultivars. Additionally, the storage method influences the antioxidant content together with the cultivar type and the harvest location.

To preserve the fruit's bioactive compounds, it is important to avoid its natural ripening and senescence, with post-harvest treatments. A post-harvest treatment capable of delaying the ripening of the cherries and enhancing their bioactive compounds and antioxidant properties was proposed by Valero et al. (Valero et al. 2011). Hence, salicylic acid, acetylsalicylic acid or oxalic acid were used in sweet cherries *Cristalina* and *Prime Giant* cultivars at a concentration of 1 mM, and then the samples were stored under cold temperature for 20 days. When compared to the control samples (non-treated cherries), it was observed that the antioxidant activity of the treated fruits lasted longer and increased while stored, as opposed to the control samples which antioxidant activity increased during the first ten days, decreasing after this time. Also, the contents in bioactive compounds were maintained for a longer period of time.

Ballistreti et al. (Ballistreri et al. 2013), evaluated the relevance of the bioactive compounds contained in sweet cherries, namely phenolic compounds and the antioxidant properties on more than twenty sweet cherries cultivars from Sicily, Italy. By performing the oxygen radical absorbance capacity (ORAC) assay, the authors were able to demonstrate that all genotypes of the sweet cherries samples possessed high antioxidant activity. Moreover, they were able to establish a correlation between the main phenolic compounds (neochlorogenic, followed by *p*-cumaroylquinic and chlorogenic acids) and the total antioxidant capacity of the analyzed cherries.

In 2015, Bastos et al. (Bastos et al. 2015), evaluated extracts, infusions and decoctions using sweet cherries stems to determine antioxidant activity (DPPH assay) and other chemical characteristics, such as lipophilic compounds, tocopherols, phenolic compounds composition, amongst others. The authors discovered that the antioxidant content of the stems was higher than in the fruit extracts, which can be explained by the presence of higher contents of phenolic acids and flavonoids in this part of the fruit.

On a recent study, Zhao et al. (Zhao et al. 2019), performed a thorough assessment on the effect that the storage temperature had on the quality and the antioxidant metabolism of the sweet cherries. Fruits were stored at 0°C and 5°C and samples were changed every 20 days, when they turned visibly rot. As main conclusions and regarding the antioxidant evaluation, when sweet cherries are stored at these temperatures, the ripening and agedness were delayed. Moreover, when fruits were stored at 0°C, they could be stored for twenty more additional days. Therefore, reactive oxygen species (ROS) production and oxidative stress were also delayed. Such results are promising as a storage method to improve quality of cherries, since oxidative stress is decreased.

Research has been focusing mainly on the fruit itself; however, in 2015 Bastos et al. (Bastos et al. 2015), assessed the stems of the plant and most recently Dziadek et al. (Dziadek, Kopeć, and Tabaszewska 2019), evaluated the antioxidant activity and determined the bioactive compounds of the leaves and petioles. The antioxidant activity was determined by three assays (ABTS, DPPH, and FRAP). Authors found that both leaves and petioles present higher antioxidant activity than the fruit itself.

Overall, the antioxidant activity in sweet cherries is due to the presence of phenolic compounds, ascorbic acid, vitamins A, C and E and carotenoids (Krinsky and Yeum 2003; Scalbert et al. 2005).

Neuroprotective

In a way, neurodegenerative diseases are ultimately related with oxidative stress, which leads to neural cells degradation and consequently to ataxia or dementia (Z. Zhang et al. 2015; Uttara et al. 2009). Literature is somewhat scarce when searching for "anti-neurodegeneration properties" and "*Prunus avium L.*" However, a few articles depict the importance of sweet cherry consumption and its impact in neurodegeneration.

Vinitha et al. (Vinitha et al. 2014), performed a study in diabetic mice in which neurotoxicity was induced by streptozotocin. These mice possessed cholinergic insufficiency, as well as low activity of choline acetyltransferase and high acetylcholinesterase activity. The animals were treated for 28 days with ethanolic cherry extracts (200 mg/kg and 400 mg/kg, oral administration), and results showed that the dried ethanolic sweet cherries extracts were capable of preventing neurotoxicity in mice, by diminishing acetylcholinesterase levels, and improving mice's memory and augmenting the levels of antioxidants. These results suggest that sweet cherries constituents are able to restrict acetylcholinesterase activity, possibly because of their antioxidant capacity. In a human based study performed on elderly adults with mild-to-moderate dementia, and to whom was given a daily intake (200 mL/day) of anthocyanin-rich sweet cherry juice, Kent et al. (Kent et al. 2017) were able to observe enhanced verbal fluency, as well as enhancement of both short- and long-term memory, when compared to the control population.

Evidences also demonstrate that phenolic compounds in sweet cherries are responsible for the neurological positive effects, since these compounds manage to cross the blood brain barrier, scavenging harmful molecules, such as ROS and reactive nitrogen species (RNS). In addition, they can chelate transition metal ions, and consequently lower neuronal damage and losses induced by neurotoxins and neuroinflammation (Daniele Del Rio et al. 2013). Moreover, polyphenols are able to reduce the production of molecules that can cross cell membrane's and generate hydroxyl groups that lead to neurodegeneration, lipid peroxidation and cell death (Kim et al. 2005). Other mechanisms of action include the inhibition of cyclooxygenases activity, and boosting of guanosine triphosphatase activity, which lead to the release of the neurotransmitter dopamine, as well as avoiding the deficit of sensitivity in Purkinje cells, which lead to the improvement of cognitive performance, as well as preventing neural cell damage (Subash et al. 2014).

Anticancer

The phenolic content, together with other components such as fiber, vitamins and carotenoids, give to *Prunus avium* L. fruit an antioxidant character, helping to prevent some diseases, namely cancer (Commisso et al. 2017; Ferretti et al. 2010; McCune et al. 2011). Phenolic compounds have been of great interest for their important anticancer role (Li et al. 2014). Epidemiological studies have shown that daily intake of phenolic-rich foods prevents some cancers, particularly those of the gastrointestinal tract such as mouth, stomach, duodenum and colon cancers, but also others such as liver, lung, gland breast or skin cancer (Li et al. 2014; Haminiuk et al. 2012). Plant and fruit phytochemicals can also inhibit the genotoxicity of some carcinogenic potential compounds from food, such as polycyclic aromatic hydrocarbons and heterocyclic aromatic amines (Platt et al. 2010). In a study by Platt et al. (Platt et al. 2010) it was shown that sweet cherry juice has an inhibitory effect on the genotoxicity of 2-amino-3-methylimidazo [4,5-*f*] quinoline.

Several *in vitro* studies have also been conducted to understand the anticancer effects related to the consumption of *Prunus avium* L. fruit. Wang and Stoner (L. S. Wang and Stoner 2008) used leukemic, keratinocyte, liver, colon, endothelial, and breast cell lines to investigate cancer-related effects of anthocyanins. These compounds have been found to have chemopreventive activities in cancer since they stimulate the expression of the enzymes glutathione reductase (GR), glutathione peroxidase (GPX) and NAD(P)H:quinone reductase, responsible for phase II detoxification reactions. In addition, anthocyanins also induce apoptosis, reduce cell proliferation and inhibit mutagenesis caused by environmental toxins (Ferretti et al. 2010; L. S. Wang and Stoner 2008). In other studies, performed on cancer cell lines it was found that after exposure to sweet cherry anthocyanins, cell cycle arrest and apoptosis induction occurred (Lazzè et al. 2004; Shih, Yeh, and Yen 2005). Other studies indicate that this cyanidin-induced cell cycle arrest is due to the inhibitory effect of this compound on the epidermal growth factor receptor (Meiers et al. 2001). It has also been reported that cyanidin reduces the risk of malignant transformation as it promotes cell differentiation (McCune et al. 2011). In addition to the above, several other studies have been performed, demonstrating the anticancer and antiproliferative properties of *Prunus avium* L. fruit. Serra et al. (Serra et al. 2011) developed a study evaluating the antiproliferative effect of nine sweet cherry cultivars on stomach (MKN45) and colon (HT29) derived human cancer cells, and it was found that *Ulster* and *Lapin* cultivars showed a higher inhibition in those cell lines. The same research group developed another study where they tested the antiproliferative activity of sweet cherry, using extracts with different mixtures of CO_2 and ethanol (Serra et al. 2010). The extract obtained with CO_2:EtOH (90:10, v/v) was found to have a higher antioxidant activity and was the most effective in inhibiting the proliferation of human cancer cells (Serra et al. 2010). It was also possible to verify that perilyl alcohol proved to be one of the main responsible for these properties (Serra et al. 2010). In 2014, Pacifico et al. (Pacifico et al. 2014) tested the potential inhibitory effects of extracts of two sweet cherry cultivars on cell growth of HepG2, A549, HeLa, SK-B-NE(2)-C and SH-SY5Y cell lines, however only one extract showed dose-dependent inhibitory activity in the HeLa cell line (Pacifico et al. 2014). More recently, Silva et al. (Silva et al. 2019) developed a study that evaluated the anticancer properties of *Prunus avium* L. fruit extract in neoplastic (PC3 and LNCaP) and non-neoplastic (PNT1A) prostate cells. These cell lines were treated with sweet cherry extracts for 48h and 96h, and a decreased cell viability was observed in all cell lines. In LNCaP cells there was also a reduction in oxidative damage and a suppression in glycolytic metabolism (Silva et al. 2019).

Currently, the forms of cancer treatment consist in the use of radioprotective and radiosensitizing agents. Thus, studies have also been carried out to understand the role of *Prunus avium* L. fruit in cancer treatment (Piccolella et al. 2018). Recently, Piccolella et al. (Piccolella et al. 2018) studied the radiomodulatory properties of a sweet cherry extract, as well as its ability to modify the response of a neuroblastoma cell line after exposure to ionizing radiation. It was found that at doses between 25 and 50 µg/mL sweet cherry extract acted as a radioprotector. However, for doses between 400 and 500 µg/mL sweet cherry extract increased the cytotoxic effects of radiation (Piccolella et al. 2018).

However, other studies show the opposite. Kulisic-Bilusic et al. (Kulisic-Bilusic et al. 2009) developed a study in which they evaluated the antioxidant activity, kappa B nuclear factor activation and the effect on cell viability in a human colon cancer cell line of various fruit juice, namely sweet cherry juice. The results indicated that although sweet cherry juice had a strong antioxidant activity, it did not have cytotoxic effects on human colon cancer cell line or nuclear factor kappa B inhibitory activity (Kulisic-Bilusic et al. 2009). Thus, this study showed no correlation between antioxidant activity and cell viability and nuclear factor kappa B inhibitory activity (Kulisic-Bilusic et al. 2009).

Anti-Inflammatory

Some of the components present in sweet cherries, namely melatonin, carotenoids, vitamins E and C, flavonoids and anthocyanins, have been related to the antioxidant and anti-inflammatory properties conferred by this fruit (Kelley, Adkins, and Laugero 2018). These properties are related to the chemical structures of polyphenols, since the aromatic structure and the presence of hydroxyl groups make polyphenols good donors of electrons or hydrogen atoms, neutralizing ROS (H. Zhang and Tsao 2016). Several studies have been developed to demonstrate the anti-inflammatory properties of *Prunus avium* L. fruit.

Two studies developed by Kelly et al. (Kelley et al. 2013, 2006) showed that ingesting 280 g of sweet cherry per day for four weeks inhibited the activity of cycloxygenases I and II (Cox I and Cox II), leading to the conclusion that there is a decrease in the inflammatory pathways. Also, in the same study it was possible to verify that the consumption of sweet cherry decreases inflammation biomarkers, namely C-reactive protein (CRP), interleukin-18 (IL-18), interleukin-1 receptor agonist (IL-1Ra), ferritin, tumour necrosis factor alpha (TNF-α), endothelin-1 (ET-1), plasminogen activator inhibitor-1 (PAI-1) and extracellular ligand of the advanced glycation end products receptor (EN-RAGE) (Kelley et al. 2013, 2006). According to the same authors, the consumption of sweet cherry helps reducing the risk of arthritis (indicated by PCR, IL-18, TNFα and IL-1Ra), among other diseases (Kelley et al. 2013). Also, Seernam et al. (Navindra P Seeram, Zhang, and Nair 2003) developed an *in vitro* study where they evaluated the activity of COX-I and COX-II. The authors demonstrated that the compounds present in sweet cherry, such as anthocyanin, cyanidin and malvidin, had an inhibitory effect on these enzymes (Navindra P Seeram, Zhang, and Nair 2003). Another study, in which the anti-inflammatory effects of cyanidin, anthocyanins, were evaluated separately, showed that sweet cherry inhibited COX-I and COX-II 28% and 47%, respectively (N. P. Seeram et al. 2001). Also, the anti-inflammatory response to common anti-inflammatory drugs was evaluated, with sweet cherry having a COX-II inhibition 5% higher than the tested drugs (N. P. Seeram et al. 2001). Another study by Hou et al. (Hou et al. 2005) also demonstrated that anthocyanins present in sweet cherry have an inhibitory effect on COX-II. These inhibitory effects were also found to be related to inhibition of mitogen-activated protein kinase (MAPK) (Hou et al. 2005).

Jacob et al. (Jacob et al. 2003) developed a study where they administered a single 280 g dose of sweet cherry to healthy women. The results showed that there was a decrease in serum urate levels, responsible for diseases such as gout (Jacob et al. 2003). An observational study by Zhang et al. (Y. Zhang et al. 2012) corroborated the results obtained by Jacob et al. (Jacob et al. 2003), since it allowed to establish a correction between sweet cherry consumption and reduction of gout disease.

Plasma nitric oxide (NO) increase is associated to several inflammatory diseases such as rheumatoid arthritis, systemic lupus erythematosus and osteoarthritis (Ferretti et al. 2010). Studies have also indicated that the consumption of anthocyanins, present in sweet cherry, inhibits the production of pro-inflammatory factors, namely NO, TNF-α and activated macrophages (Ferretti et al. 2010; Jian Wang and Mazza 2002; J. Wang and Mazza 2002). He et al. (He et al. 2006) developed an *in vivo* study using an induced arthritis model where they evaluated the anti-inflammatory response. For this, the rats were daily fed with 10, 20 or 40 mg/kg of anthocyanins present in sweet cherry for 28 days. Levels of TNF-α and prostaglandin E2 (PGE2) were evaluated and a significant reduction was observed for the highest dose (He et al. 2006). Another *in vivo* study was developed by Wu et al. (Wu et al. 2016) to evaluate the role of sweet cherry anthocyanin in oxidative stress and inflammation associated with obesity in mice fed with a high-fat diet. For this, sweet cherry anthocyanin was added to the daily diet of mice for eight weeks (Wu et al. 2016). There was a reduction in lipid profile and serum glucose and leptin levels, as well as in MDA production. It was also possible to verify an increase in GPX and superoxide dismutase (SOD) activities, demonstrating the potential of sweet cherry in reducing oxidative stress and obesity-induced inflammation (Wu et al. 2016).

Antimicrobial

As described above, sweet cherries are a rich source of flavonoids and polyphenols (Kim et al. 2005; Mikulic-Petkovsek et al. 2016; Ferretti et al. 2010). In plants, these compounds are increasingly believed to aid the physiological survival of plants, protecting them against pathogens (bacterial or fungal) and UV radiation (Nowak et al. 2016). Polyphenols of plant origin have been reported to have a variety of biological effects, including among others antimicrobial activity (Brglez Mojzer et al. 2016). Specifically, some phenolic compounds, such as resveratrol, hydroxytyrosol, quercetin, and several phenolic acids, have been reported to inhibit various pathogenic microorganisms (Luís et al. 2014). Flavonoids are another group of phenolic compounds, which are known to be synthesized by plants in response to microbial infection.

Regarding the antimicrobial properties of sweet cherries, there has been a backlog of information that, even though well documented, it certainly demands scrutiny and analysis, as it was also mentioned previously by other authors (L. B. Hanbali et al. 2013).

It was shown that the juice and both aqueous and methanolic extracts of sweet cherries were able to inhibit the growth of *Staphylococcus aureus*, *Staphylococcus epidermidis*, *Streptococcus pyogenes* and *Cutibacterium acnes* (formerly *Propionibacterium acnes*) with values of Minimum Inhibitory Concentration (MIC) ranging from 6.13 to 15.6 mg/mL (Ördögh et al. 2010). These results suggest the potential application of sweet cherries or even the by-products of food industry, namely those from the juice production, to the development of natural drugs or cosmetics to treat acnes vulgaris (Ördögh et al. 2010). Furthermore, another work that evaluated the antimicrobial activity of sweet cherry juice against a broad of Gram-positive and Gram-negative pathogenic bacteria demonstrated the measurable attenuating effect of *P. avium* extracts on the differential growth bacterial strains, with a clear pattern of sensitivity/resistance across the descending concentrations of extracts spectrum (L. Hanbali et al. 2013).

In a study aiming the evaluation of the phenolic composition in the early and late ripening of sweet cherries, together with the assessment of their activity against *Alternaria alternata* (one of the main pathogens of sweet cherry that produces specific mycotoxins such as tenuazonic acid and alternariols) it was found that conjugated phenolics of the late ripening cultivars achieved the highest antifungal activity and almost completely inhibited *A. alternata* and tenuazonic acid production, probably due to their high levels of lipophilic phenols and stronger antioxidant activities (Wang et al. 2017).

It was also demonstrated that a fermented sweet cherry juice was able to inhibit the ulcer linked pathogen *Helicobacter pylori* without affecting the beneficial lactic acid bacteria *Bifidobacterium longum* (Ankolekar et al. 2011). Investigations into the mechanism of action against *H. pylori* revealed that one of the likely mechanisms of action was by inhibition of proline oxidation via proline dehydrogenase (Ankolekar et al. 2011).

Further studies will be needed to provide clear results about the antimicrobial properties of sweet cherries, because the great amount of sugars present in this fruit may affect the antimicrobial activity together with the fact that polyphenols are mainly present in their glycosylated forms.

Antidiabetic and Cardiovascular

Diabetes mellitus is a metabolic disease, featured by chronic hyperglycaemia and glucose intolerance, affecting millions of people worldwide. Although pharmacotherapy is available, its limitation for the control of blood glucose has boosted the phytotherapy research for diabetes management (Saleh, El-Darra, and Raafat 2019). As previously stated, sweet cherries are rich in polyphenols including hydrocinnamic acids, anthocyanins, flavonols, and flavan-3-ols. However, anthocyanins have been the centre of attention regarding antidiabetic properties. These specific compounds are suggested to reduce blood glucose, glucosuria and glycated haemoglobin (Hb A1c), prevent free radical production, increase insulin secretion and improve insulin resistance (Sancho and Pastore 2012; Ataie-Jafari et al. 2008). Anthocyanins are secondary metabolites of higher plants, water-soluble poly-hydroxy and methoxy poly-glycosides derived from 2-phenylbenzopyrylium (flavylium cation), and are known for providing the blue, red and purple colours of these fruits (Sancho and Pastore 2012).

Overall, few *in vitro* studies have been conducted, revealing anthocyanins anti-diabetic effects and action on α-glucosidase and pancreatic α-amylase, their major targets (Belwal et al. 2017). Cao et al. (Cao et al. 2015) observed that sweet cherry phenolics-rich extract was effective in promoting HepG2 cell line glucose consumption. However, the authors differentiated phenolics subclasses into an anthocyanin-rich fraction (ARF), hydrocinnamic acid-rich fraction (HRF), and flavonol-rich fraction (FRF) and found that the 3 fractions promoted HepG2 glucose consumption to different levels. The promotion effects of HRF and FRF appeared to be stronger than the ARF. Gonçalves et al. (A. C. Gonçalves et al. 2018) also reported the inhibition of α-glucosidase activity by sweet cherry extracts, obtaining much lower half maximal inhibitory concentration (IC_{50}) values when compared to the positive control, acarbose, a common antidiabetic drug taken orally.

Lachin and Reza (Lachin and Reza 2012) evaluated the effect of an ethanolic extract of cherry fruit on alloxan induced diabetic rats. For this study, the authors divided 36 male wistar rats into 6 groups and diabetes was induced by intraperitoneal injection of 120 mg/kg alloxan. These authors observed a significant reduction on blood glucose and urinary microalbumin as well as an increased creatinine secretion level in urea associated in rats treated with sweet cherries extracts. The authors concluded that cherries appear to be helpful in the control of diabetes and its complications. Lachin (Lachin 2014) evaluated the antioxidant effect of extracts from cherries on the blood glucose and urinary creatinine and microalbumin level, also in alloxan-induced diabetic rats. Lachin (Lachin 2014) shown that the cherry extract had a superior anti-diabetic activity appearing to be an excellent antidiabetic candidate.

There are currently more studies on the influence of phenolic compounds, mainly anthocyanins, in the prevention of diabetes. Unfortunately, there are not studies involving sweet cherries nor other varieties of cherries (eg. *Prunus cerasus*, Cornelian cherry, etc.). Some of these studies will be discussed in the following lines.

Jayaprakasam et al. (Jayaprakasam et al. 2005) determined the ability of several anthocyanins and anthocyanidins to stimulate insulin secretion from rodent pancreatic β-cells in the presence of 4 and 10 mM glucose concentrations. On the other hand, Nowicka et al. (Nowicka, Wojdyło, and Samoticha 2016) screened *Prunus* smoothies to determine the polyphenols profile and find correlation between the polyphenols and antioxidant capacity, as well as α-glucosidase and α-amylase inhibitory effect. The results have shown that anthocyanins and flavonols were the compounds with greatest impact on the antioxidant capacity and anti-α-glucosidase activity, while the flavan-3-ols could be responsible for α-amylase inhibition (Nowicka, Wojdyło, and Samoticha 2016). The authors suggested that *Prunus* smoothies are natural sources of polyphenolic antioxidants with great potential to control the early stages of post-prandial hyperglycaemia (Nowicka, Wojdyło, and Samoticha 2016).

Additionally to *in vitro* studies, some molecular modelling investigations were performed by Homoki et al. (Homoki et al. 2016), which helped predicting the *in vivo* reactions. These authors evaluated the anthocyanin composition, antioxidant efficiency, and α-amylase inhibitor activity of different Hungarian cherry cultivars. A high antioxidant capacity and anthocyanin content were observed for two cultivars, with cyanidin-3-rutinoside revealing as the main anthocyanin of the cherry samples. In addition, the cherry extracts and the anthocyanins competitively inhibited the human serum albumin (HSA) catalysed hydrolysis. The inhibitors binding to the HSA active site were investigated by molecular modelling, after which they were compared to that of acarbose, and the main docking poses of cyanidin derivatives inhibiting HSA were found similar to the interactions of acarbose at the proposed subsite (Homoki et al. 2016).

Regarding *in vivo* studies, a vast number has been conducted in order to justify the protective activities of sweet cherry compounds against insulin resistance diabetes (Belwal et al. 2017). Jayaprakasam et al. (Jayaprakasam et al. 2006) reported an amelioration of obesity and glucose intolerance in high-fat-fed C57BL/6 mice by components of cornelian cherry. The authors found that C57BL/6 mice fed with the high-fat diet containing anthocyanins and ursolic acid resulted in a normalized glucose intolerance, preserved islet architecture and an extremely elevated level of circulating insulin. For this reason, it was suggested that consumption of cornelian cherry or other fruits containing anthocyanins and ursolic acid may lead to the risk reduction of diabetes. Seymor et al. (E Mitchell Seymour et al. 2008) observed an altered hyperlipidaemia, hepatic steatosis, and hepatic peroxisome proliferator-activated receptors in rats treated with tart cherry. The study was conducted in Dahl Salt-Sensitive rats with insulin resistance and hyperlipidaemia. For 90 days, Dahl rats were supplemented with an intake of either 1% (wt:wt) freeze-dried whole tart cherry or with 0.85% additional carbohydrate. After the period of the study, a reduced fasting blood glucose, hyperlipidaemia, hyperinsulinemia, and reduced fatty liver was associated to the cherry-enriched diet. Additionally, in the same study, an association was observed between the cherry diet and enhanced hepatic tissue peroxisome proliferator-activated receptors (PPARs), namely, PPAR-α mRNA and hepatic PPAR-α target acyl-coenzyme A oxidase mRNA. The authors also observed an increased plasma antioxidant capacity, suggesting a risk reduction for metabolic syndrome and type 2 diabetes when there is a relevant tart cherry consumption (E Mitchell Seymour et al. 2008). Later, Seymor et al. (E. M. Seymour et al. 2009) confirmed that tart cherry-enriched diets are associated with a significant reduction of body weight, abdominal fat, blood lipids, plasma inflammation, and fasting glucose. The authors have proven, once again, the association between tart cherry consumption and the increased expression of PPAR isoforms and related genes in abdominal fat, reduced abdominal fat IL-6 and TNF-α, and reduced factor nuclear kappa B (NFkB) activity (E. M. Seymour et al. 2009).

Other *in vivo* study was done by Asgary et al. (Asgary et al. 2014), that performed a biochemical and histopathological study of the anti-hyperglycaemic and anti-hyperlipidaemic effects of cornelian cherry in alloxan-induced diabetic rats. The authors observed that the effects of cornelian cherry were comparable to those of glibenclamide, an oral hypoglycaemic agent, at the tested doses. Additionally, the histopathological examinations revealed less severe hepatic portal inflammation in the cornelian cherry treated when compared to the other study groups (Asgary et al. 2014). Dzydzan et al. (Dzydzan et al. 2019) evaluated the antidiabetic effects of extracts of fruits of cornelian cherries on rats with streptozotocin-induced diabetes mellitus and observed that the diet supplementation with the extracts significantly decreased the concentration of glucose in the diabetic animals. The authors also studied the oral glucose tolerance tests, observing that blood glucose concentration peaked 30 min after oral glucose administration, and decreased to baseline at 60 min in the control rats. Though, blood glucose concentrations were still higher in diabetic rats at 60 min after glucose ingestion, the blood glucose levels peaked on 30 min and then slowly decreased to the fasting level after intake of cornelian cherry extracts. Statistically significant differences were observed after 120 min between cornelian cherry-treated groups and the diabetic group (Dzydzan et al. 2019).

Apart from the latter *in vivo* studies, clinical studies become crucial in order to consolidate the cherries beneficial effects. Nevertheless, few have been conducted on their effect against insulin resistance under diabetes and/or obesity conditions (Kelley, Adkins, and Laugero 2018). Garrido et al. (M Garrido et al. 2013) analysed the nutritional and functional properties of a *Jerte Valley* cherry for its potential antioxidant status improvement in humans. After a placebo-controlled trial with human volunteers, the ingestion of this product seemed to improve their antioxidant status. Kelley et al. (Kelley et al. 2006) aimed to determine the effects of sweet cherries consumption on plasma lipids and markers of inflammation in healthy humans. The authors observed that cherry consumption did not affect the plasma lipids and markers of inflammation, and also did not affect fasting blood glucose or insulin concentrations or a number of other chemical and haematological variables (Kelley et al. 2006).

Moreover, evidence exists suggesting that cherry consumption may promote healthy glucose regulation. The outcomes of studies performed in human, animal, and cells point that anthocyanins may decrease blood glucose and hepatic glucose output (Kelley, Adkins, and Laugero 2018). Also, there is proof of a decrease in the production of glucagon by pancreatic α-cells, and an increase of hepatic glucose uptake as well as production of insulin by pancreatic β-cells (Kelley, Adkins, and Laugero 2018; Crepaldi et al. 2007).

It would be beneficial if cherry constituents could regulate blood glucose levels or induce insulin production by pancreatic β-cells in diabetes. However, there is still little clinical evidence and future studies are needed to confirm whether these findings translate to reduced risk of this epidemic metabolic disease (Kelley, Adkins, and Laugero 2018; Ghosh and Konishi 2007).

Consumption of sweet cherries or tart cherry concentrate by healthy adults did not modify the concentrations of blood lipids, including triglycerides (TG), low-density-lipoproteins (LDL), very-low-density lipoproteins (VLDL), high-density lipoproteins (HDL), total cholesterol, number of different lipoprotein particles and their sizes in healthy adults (Kelley et al. 2013; Lynn et al. 2014). In contrast to the studies involving healthy participants, another study with overweight and obese subjects presenting elevated blood lipids reported a decrease in VLDL and TG/HDL ratio following consumption of tart cherry juice for 4 weeks (Martin et al. 2011). It seems that the lipid profile of study participants prior to the supplementation with tart cherries (Martin et al. 2011) versus sweet cherries (Kelley et al. 2006) rather than the type of cherries may have contributed to the different results between these two studies. As stated above, cherry extracts and purified anthocyanins decreased liver triglycerides and cholesterol in mouse and rat models and prevented the high fat diet induced development of NAFLD (E Mitchell Seymour et al. 2008; Snyder et al. 2016). The decrease in blood pressure caused by the prolonged consumption of cherries may have resulted from the decrease in endothelin-1 (ET-1), one of the most potent vasoconstrictors (Kelley et al. 2013). Nitrous oxyde produced by endothelial NO synthase (eNOS) is an important vasodilator, and its expression was increased by the addition of cyanidin-3-glucoside to cultured human umbilical vein endothelial cells and bovine vascular endothelial cells (Edwards et al. 2015). Hence, altered expression of both ET-1 and eNOS by cherry consumption may have contributed to the decrease in blood pressure.

Other Health Benefits

Kelley et al. (Kelley, Adkins, and Laugero 2018) published in 2018 an excellent review on clinical studies involving consumption of cherries and their products. Among the health proprieties previously described, these authors have gathered a series of clinical trials from which it is possible to verify the effects of cherries on exercise-induced muscle damage and recovery. In eight of the nine studies, pain and loss of strength were significantly reduced by cherry consumption. The attenuation of exercise-induced muscle damage by cherries seems to be related to the antioxidant and anti-inflammatory properties of anthocyanins and other phenolic compounds found in cherries (Coelho, Lima, and Prestes 2015). Unfortunately, these studies have been conducted using other varieties of cherries, namely *Prunus cerasus* L. and not *Prunus avium* L.

The efficacy of these cherries on sleep, mood, and cognitive functions was verified. Both quality and quantity of sleep were improved by the consumption of sweet cherries (M Garrido et al. 2013; María Garrido et al. 2009). Effect on sleep could be detected within 3 days of consuming sweet cherries (141 g or 25 cherries/day). The studies using sweet cherries also reported a decrease in urinary cortisol and anxiety, and improved mood (M Garrido et al. 2013; María Garrido et al. 2012).

ANALYTICAL METHODS TO DETERMINE THE COMPOSITION OF *PRUNUS AVIUM* L.

Phenolic compounds have gained more relevance to scientists because of the several beneficial effects on human health (Mikulic-Petkovsek et al. 2016; J. Gonçalves, Ramos, Luís, et al. 2019; Kelley, Adkins, and Laugero 2018). In fact, there are several methods to evaluate the composition of sweet cherries, especially the amount of total phenolic compounds and serotonin, among other nutrients (Rosado, Henriques, et al. 2017; Rosado, Barroso, et al. 2017; Ignat, Volf, and Popa 2013; Antolovich et al. 2000). However, in order to determine and quantify the composition in sweeet cherries in terms of phenolic acids, flavonoids and anthocyanins, extraction processes and analytical methods capable of identifying them are required.

The chromatographic technique of choice for studying the composition of this fruit is liquid chromatography coupled to mass spectrometry or diode array detectors (LC-MS/DAD). The freeze-drying process of the sweet cherry samples is also considered a common procedure for these determinations. Some of the most recent work, which identifies the major compounds present in sweet cherries is presented below.

Martini et al. (Martini, Conte, and Tagliazucchi 2017) developed a methodology to evaluate the profile of six cultivars of sweet cherries (*Della Marca*, *Celeste*, *Bigarreau*, *Durone Nero*, *Lapins* and *Moretta*) from Vignola, Italy. After harvesting, the cherries were either processed or frozen and stored at -80°C. To extract the phenolic compounds, 30 g of cherries were homogenized with 50 mL of water/methanol/formic acid (28:70:2, v/v/v), the suspension was mixed for 120 min at 30°C, centrifuged and the supernatant was filtered. Subsequently, the extracts (1 mL each) were subjected to solid-phase extraction through preconditioning with 4 mL of acidified methanol containing 0.1% of formic acid, followed by 5 mL of acidified water with 0.1% formic acid. The samples were then eluted with 6 mL of acidified water and the phenolic compounds were desorbed with 3 mL of acidified methanol and further analysed. The phenolic compounds were identified and quantified by liquid chromatography ion trap mass spectrometer. The developed method identified 86 compounds among hydroxycinnamic acids, hydroxybenzoic acids, anthocyanins, flavan-3-ols, flavonols and other flavonoids. The authors identified hydroxycinnamic acid derivatives as the main represented class of phenolic compounds, with the exception of *Lapins* and *Durone Marca* cultivars, where the main class of phenolic compounds were anthocyanins and flavano-3-ols, respectively. They also concluded that this approach allows the identification of the relationship between cultivars and their composition.

Commisso et al. (Commisso et al. 2017) analyzed 18 sweet cherry cultivars (*Sandra Tardiva, Roana, Kordia, Early Bigi, Bella Italia, Burlat, Van, Durone Rosso, Sandra, Durona del Chiampo, Giorgia, Ferrovia, Lapins, Regina, Black Star, Romana, Grace Star* and *Milanese*) from Vicenza, Italy. For that, 300 mg of frozen powder of each sample was used to analyze polyphenols, by extracting with 3 volumes of cold methanol acidified with 1% hydrochloric acid. The samples were then vortexed, sonicated for 15 min and centrifuged. To quantify the ascorbic acid in fresh samples, 20 g of the samples were homogenized with 0.2% potassium metabisulfite powder. Subsequently, 1 g of these samples was diluted with 20 mM of monosodium phosphate (pH 2.14), vortexed, centrifuged, filtered and stored at 4°C in the dark. For frozen fruit samples, 150 mg of powder was extracted with nine volumes of 0.02% potassium metabisulfite in 20 mM sodium dihydrogen phosphate buffer (pH 2.14) and the remaining process being the same as described above for fresh fruit. All the extracts were diluted (1:3) and analyzed by HPLC-MS, revealing the presence of the procyanidins, flavonoids, hydroxycinnamic acids, catechin and epicatechin, pelargonidin rutinoside and peonidin rutinoside. Analysig the extracts by HPLC-DAD, hydroxycinnamic acids, flavonols, anthocyanins, neochlorogenic, chlorogenic, *p*-coumaroyl quinic and ascorbic acids were identified. The authors concluded that the composition of sweet cherries varies with the environment, climatic factors, genotype as well as agronomic practices. They also concluded that each cultivar has a specific phenolic composition, depending on each cultivar, while the edaphoclimatic conditions may have a less significant impact on the secondary metabolism of sweet cherries.

Piccolella et al. (Piccolella et al. 2018) evaluated the sweet cherry fruits of the cultivar *Della Recca* from Caserta, Italy. The cherries were cleaned, frozen, lyophilized and then milled with liquid nitrogen. Aliquots of 5 g of cherries underwent ultrasound assisted extraction with 150 mL of water as extraction solvent, a process repeated for three cycles of 30 min each. After centrifugation, the supernatants were combined, dried using a rotary evaporator and the final product was considered the crude extract. To study the metabolic composition of the aqueous extract, aliquots of 0.7 g of cherries were chromatographically analyzed by an Amberlite XAD-4, being some of the samples eluted with water for and others with methanol. Subsequently, the samples were analyzed by UV-Vis, which suggested the existence of polyphenols such as flavonoids and/or phenolic acids. To clarify the chemical composition and to quantify the compounds, samples were analyzed by HPLC-MS. This technique showed to be efficient in revealing compounds present in minor concentrations, identifying and quantifying 14 secondary metabolites namely chlorogenic acid and flavonoids, together with carbohydrates and polyols.

More recently, Gonçalves et al. (J. Gonçalves, Ramos, Rosado, et al. 2019) developed and validated a method to quantify 9 phenolic compounds in six different sweet cherry cultivars (*Saco, Summit, Sweetheart, Primulat, Brooks and Earlise*) from Fundão, Portugal. For that, 200 g of each sweet cherry, fresh or thawed, was lyophilized. Successively, 20 mL of methanol acidified with 0.1% hydrochloric acid were added to each sample of freeze-dried fruit and the mixture was kept in a thermostatic bath during 2 h at 35°C. This extraction was carried out twice and the extract obtained was centrifuged, evaporated to dryness and injected into the chromatographic system. The phenolic compounds of the samples were quantified by HPLC-DAD. In this study, cyanidin, cyanidin-3-*O*-glucoside, quercetin, gallic acid, *p*-coumaric acid, rutin, epicatechin, quercetin-3,4′-di-*O*-glycoside and chlorogenic acid were detected. The authors verified the absence of quercetin and cyanidine and concluded that the predominant compound was quercetin-3-4′-di-*O*-glycoside, being cyanidine-3-*O*-glycoside the most frequent compound in sweet cherries. They also observed that the *Saco* cultivar had the highest content of polyphenols.

Also in the same year, Gao et al. (Y. Gao et al. 2019) developed a methodology able to quantify phenolic compounds in 5 cultivars of *Prunus avium* L. (*Caihong, Sunburst, Summit, 23-51* and *Valeri*) from Beijing (China) by ultra-performance liquid chromatography coupled with tandem mass spectrometry. To identify the phenolic compounds, 1 g of freeze-dried powder from each sweet cherry cultivar (after lyophilization) was extracted with 20 mL of methanol (80%, v/v) containing 0.5% hydrochloric acid. The resulting extract was ultrasonicated, the mixture was centrifuged and the supernatant was collected. This process was repeated three times. Subsequently, the supernatants were evaporated and extracted again three times with 15 mL diethyl ether/ethyl acetate (1:1, v/v). To analyse the flavonoids, 2 g of lyophilized powder of cherries was extracted with 30 mL of methanol with 0.5% of hydrochloric acid in the dark, the mixture was centrifuged and the supernatant was collected and filtered for analysis. The authors concluded that there is a great variability in the phenolic content within cultivars and different edible parts. The most abundant phenolic acid was neochlorogenic acid and the most abundant anthocyanin was cyanidin-3-*O*-rutinoside. They also concluded that, in general, phenolic compounds are found at a higher level in the peel compared to flesh and that vanillic and ferulic acids were the most abundant present in flesh.

Furthermore, Acero et al. (Acero et al. 2019) considered five varieties of *Prunus avium* L. (*Navalinda*, *Jarandilla*, *Pico Colorado*, *Van* and *Sunburst*) from the Jerte Valley (Spain), to analyse the total phenolics, flavonoids and anthocyanins content of ripened fruits. For the extraction process, 25 g of each sweet cherry cultivar was ground with liquid nitrogen and then extracted with 250 mL of methanol with 0.1% of hydrochloric acid. The samples were sonicated at room temperature for 10 minutes, a process that was repeated four times, then centrifuged and the supernatants were evaporated and stored at 4°C. The stored samples were dissolved in methanol by adding 1000 μL/10 mg of dried sample and were further analysed. The profiles of each cultivar were obtained using ultra high performance liquid chromatography coupled to a quadrupole-time-of-flight mass spectrometry. The main compounds present in the sweet cherry extracts were 3 hydroxycinnamic acids (3-*O*-caffeoylquinic, 4-*O*-caffeoylquinic and 3-*p*-coumarylquinic acids), 4 anthocyanins (cyanidin-3,5-*O*-dihexoside, cyanidin-3-*O*-glucoside, cyanidin-3-*O*-rutinoside and peonidin-3-*O*-rutinoside) and 2 flavonoids (isorhamnetin-*O*-hexoside and quercetin-3-*O*-rutinoside). The authors concluded that *Navalinda* and *Van* are the cultivars with the highest concentrations of total phenolic compounds, and *Van* is the one that presented the highest concentration of anthocyanins. The cherries from *Pico Colorado* had the highest flavonoid content and the lowest anthocyanin content. On the other hand, the *Sunburst* cultivar presented the lowest amounts of flavanoids and anthocyanins. The authors also concluded that the main anthocyanins of sweet cherries are cyanidin-3-*O*-glucoside and cyanidin-3-*O*-rutinoside, while peonidin-3-*O*-rutinoside and pelargonidin are present in smaller amounts. Isorhamnetin-*O*-dihexoside and quercetin-3-*O*-ruthinoside are the main flavonoids that are present in the studies sweet cherries.

In general, researchers have developed analytical methodologies to determine the chemical composition of sweet cherries, concluding that each cultivar has a unique composition, and even between the same cultivar there can be differences depending, for example, on the country, the edaphoclimatic conditions and the treatment applied to the trees (Papapetros et al. 2018, 2019).

CONCLUSION

The production and consumption of sweet cherries have been increasing due to the recognition of their health benefits by consumers. Evidence from published reports is reasonably strong to indicate that sweet cherries consumption decreases markers of oxidative stress, inflammation, exercise-induced muscle soreness and loss of strength, and blood pressure. However, the beneficial effects in arthritis, diabetes, blood lipids, sleep, cognitive functions, and possibly mood have not yet been evidenced because of the little number of studies and some inconsistencies among the results; for this reason, additional studies are needed to support these claims. One of the current major concerns is to find out what is the composition and to quantify the major bioactive compounds present in sweet cherries, especially the phenolic compounds. In this sense, analytical developments mainly associated to liquid chromatography coupled to mass spectrometry represent a huge advance in this area, since they also allow the evaluation of the presence of new molecules with potential health effects. On the other hand, future investigations are required in order to better understand the bioaccessibility and bioavailability of phenolic compounds in cherries, due to the fact that this is a limiting factor of these compounds.

ACKNOWLEDGMENTS

This work is supported by FEDER funds through the POCI - COMPETE 2020 - Operational Programme Competitiveness and Internationalisation in Axis I - Strengthening research, technological development and innovation (Project POCI-01-0145-FEDER-007491) and National Funds by FCT - Fundação para a Ciência e a Tecnologia (Project UID/Multi/00709/2019). T. Rosado acknowledges the Centro de Competências em Cloud Computing in the form of a fellowship (C4_WP2.6_M1 – Bioinformatics; Operação UBIMEDICAL – CENTRO-01-0145-FEDER-000019 – C4 – Centro de Competências em Cloud Computing, supported by Fundo Europeu de Desenvolvimento Regional (FEDER) through the Programa Operacional Regional Centro (Centro 2020). S. Soares and J. Gonçalves acknowledge the FCT in the form of fellowships (SFRH/BD/148753/2019) and (SFRH/BD/149360/2019), respectively. A. Luís acknowledges the contract of Scientific Employment in the scientific area of Microbiology financed by FCT.

REFERENCES

Acero, Nuria, Ana Gradillas, Marta Beltran, Antonia García, and Dolores Muñoz Mingarro. 2019. "Comparison of Phenolic Compounds Profile and Antioxidant Properties of Different Sweet Cherry (*Prunus Avium* L.) Varieties." *Food Chemistry* 279: 260–71. https://doi.org/10.1016/j.foodchem.2018.12.008.

Antolovich, Michael, Paul Prenzler, Kevin Robards, and Danielle Ryan. 2000. "Sample Preparation in the Determination of Phenolic Compounds in Fruits." *Analyst* 125 (January): 989–1009. https://doi.org/10.1039/b000080i.

Arakawa, Osamu. 1993. "Effect of Ultraviolet Light on Anthocyanin Synthesis in Sweet Cherry, Cv. Sato NishikLight-Colored." *J. Japan. Soc. Hort. Sci*. Vol. 62.

Asgary, Sedigheh, Mahmoud Rafieian-Kopaei, Fatemeh Shamsi, Somayeh Najafi, and Amirhossein Sahebkar. 2014. "Biochemical and Histopathological Study of the Anti-Hyperglycemic and Anti-Hyperlipidemic Effects of Cornelian Cherry (Cornus Mas L.) in Alloxan-Induced Diabetic Rats." *Journal of Complementary and Integrative Medicine* 11 (2): 63–69. https://doi.org/10.1515/jcim-2013-0022.

Ataie-Jafari, Asal, Saeed Hosseini, Farzaneh Karimi, and Mohammad Pajouhi. 2008. "Effects of Sour Cherry Juice on Blood Glucose and Some Cardiovascular Risk Factors Improvements in Diabetic Women A Pilot Study." *Nutrition and Food Science* 38 (4): 355–60. https://doi.org/10.1108/00346650810891414.

Ballistreri, Gabriele, Alberto Continella, Alessandra Gentile, Margherita Amenta, Simona Fabroni, and Paolo Rapisarda. 2013. "Fruit Quality and Bioactive Compounds Relevant to Human Health of Sweet Cherry (*Prunus avium* L.) Cultivars Grown in Italy." *Food Chemistry* 140 (4): 630–38. https://doi.org/10.1016/J.FOODCHEM.2012.11.024.

Bastos, Claudete, Lillian Barros, Montserrat Dueñas, Ricardo C. Calhelha, Maria João R. P. Queiroz, Celestino Santos-Buelga, and Isabel C.F.R. Ferreira. 2015. "Chemical Characterisation and Bioactive Properties of

Prunus avium L.: The Widely Studied Fruits and the Unexplored Stems." *Food Chemistry* 173 (April): 1045–53. https://doi.org/10.1016/j.foodchem.2014.10.145.

Belitz, H. D., W. Grosch, and P. Schieberle. 2004. "Fruits and Fruit Products." In *Food Chemistry*, 806–61. Berlin, Heidelberg: Springer Berlin Heidelberg. https://doi.org/10.1007/978-3-662-07279-0_19.

Belwal, Tarun, Seyed Fazel Nabavi, Seyed Mohammad Nabavi, and Solomon Habtemariam. 2017. "Dietary Anthocyanins and Insulin Resistance: When Food Becomes a Medicine." *Nutrients* 9 (10). https://doi.org/10.3390/nu9101111.

Bokkenheuser, V. D., C. H. L. Shackleton, and J. Winter. 1987. "Hydrolysis of Dietary Flavonoid Glycosides by Strains of Intestinal Bacteroides from Humans." *Biochemical Journal* 248 (3): 953–56. https://doi.org/10.1042/bj2480953.

Cao, Jinping, Xin Li, Yunxi Liu, Feng Leng, Xian Li, Chongde Sun, and Kunsong Chen. 2015. "Bioassay-Based Isolation and Identification of Phenolics from Sweet Cherry That Promote Active Glucose Consumption by HepG2 Cells." *Journal of Food Science* 80 (2): C234–40. https://doi.org/10.1111/1750-3841.12743.

Chockchaisawasdee, Suwimol, John B. Golding, Quan V. Vuong, Konstantinos Papoutsis, and Costas E. Stathopoulos. 2016. "Sweet Cherry: Composition, Postharvest Preservation, Processing and Trends for Its Future Use." *Trends in Food Science and Technology* 55 (September): 72–83. https://doi.org/10.1016/j.tifs.2016.07.002.

Coelho, Leonardo, Rabello De Lima, and Jonato Prestes. 2015. "*Consumption of Cherries as a Strategy to Attenuate Exercise-Induced Muscle Damage and Inflammation in Humans*" 32 (5): 1885–93. https://doi.org/10.3305/nh.2015.32.5.9709.

Commisso, Mauro, Martino Bianconi, Flavia Di Carlo, Stefania Poletti, Alessandra Bulgarini, Francesca Munari, Stefano Negri, et al. 2017. "Multi-Approach Metabolomics Analysis and Artificial Simplified Phytocomplexes Reveal Cultivar-Dependent Synergy between Polyphenols and Ascorbic Acid in Fruits of the Sweet Cherry (*Prunus avium* L.).” *PLoS ONE* 12 (7): 1–23. https://doi.org/10.1371/journal.pone.0180889.

Crepaldi, Gaetano, M. Carruba, M. Comaschi, S. Del Prato, G. Frajese, and G. Paolisso. 2007. “Dipeptidyl Peptidase 4 (DPP-4) Inhibitors and Their Role in Type 2 Diabetes Management.” *Journal of Endocrinological Investigation* 30 (7): 610–14. https://doi.org/10.1007/BF03346357.

Cubero, J., F. Toribio, M. Garrido, M. T. Hernández, J. Maynar, C. Barriga, and A. B. Rodríguez. 2010. “Assays of the Amino Acid Tryptophan in Cherries by HPLC-Fluorescence.” *Food Analytical Methods* 3 (1): 36–39. https://doi.org/10.1007/s12161-009-9084-1.

Dembitsky, Valery M., Sumitra Poovarodom, Hanna Leontowicz, Maria Leontowicz, Suchada Vearasilp, Simon Trakhtenberg, and Shela Gorinstein. 2011. “The Multiple Nutrition Properties of Some Exotic Fruits: Biological Activity and Active Metabolites.” *Food Research International*. https://doi.org/10.1016/j.foodres.2011.03.003.

Demir Taki. 2013. “D*etermination of Carotenoid, Organic Acid and Sugar Content in Some Sweet Cherry Cultivars Grown in Sakarya, Turkey*” 11 (2): 73–75. https://www.researchgate.net/publication/283815342.

Dias, M. Graça, M. Filomena G.F.C. Camões, and Luísa Oliveira. 2009. “Carotenoids in Traditional Portuguese Fruits and Vegetables.” *Food Chemistry* 113 (3): 808–15. https://doi.org/10.1016/j.foodchem.2008.08.002.

Dziadek, Kinga, Aneta Kopeć, and Małgorzata Tabaszewska. 2019. “Potential of Sweet Cherry (*Prunus avium* L.) by-Products: Bioactive Compounds and Antioxidant Activity of Leaves and Petioles.” *European Food Research and Technology* 245 (3): 763–72. https://doi.org/10.1007/s00217-018-3198-x.

Dzydzan, Olha, Ivanna Bila, Alicja Z. Kucharska, Iryna Brodyak, and Natalia Sybirna. 2019. “Antidiabetic Effects of Extracts of Red and Yellow Fruits of Cornelian Cherries (: *Cornus mas* L.) on Rats with Streptozotocin-Induced Diabetes Mellitus.” *Food and Function* 10 (10): 6459–72. https://doi.org/10.1039/c9fo00515c.

Edwards, Michael, Charles Czank, Gary M Woodward, Aedín Cassidy, and Colin D Kay. 2015. "Phenolic Metabolites of Anthocyanins Modulate Mechanisms of Endothelial Function." *Journal of Agricultural and Food Chemistry* 63 (9): 2423–31. https://doi.org/10.1021/jf5041993.

Faniadis, D., P. D. Drogoudi, and M. Vasilakakis. 2010. "Effects of Cultivar, Orchard Elevation, and Storage on Fruit Quality Characters of Sweet Cherry (*Prunus avium* L.)." *Scientia Horticulturae* 125 (3): 301–4. https://doi.org/10.1016/J.SCIENTA.2010.04.013.

Fazzari, Marco, Lana Fukumoto, Giuseppe Mazza, Maria A. Livrea, Luisa Tesoriere, and Luigi Di Marco. 2008. "In Vitro Bioavailability of Phenolic Compounds from Five Cultivars of Frozen Sweet Cherries (*Prunus avium* L.)." *Journal of Agricultural and Food Chemistry* 56 (10): 3561–68. https://doi.org/10.1021/jf073506a.

Ferretti, Gianna, Tiziana Bacchetti, Alberto Belleggia, and Davide Neri. 2010. "Cherry Antioxidants: From Farm to Table." *Molecules* 15 (10): 6993–7005. https://doi.org/10.3390/molecules15106993.

Fürstenberg-Hägg, Joel, Mika Zagrobelny, and Søren Bak. 2013. "Plant Defense against Insect Herbivores." *International Journal of Molecular Sciences*. MDPI AG. https://doi.org/10.3390/ijms140510242.

Gao, L., and G. Mazza. 1995. "Characterization, Quantitation, and Distribution of Anthocyanins and Colorless Phenolics in Sweet Cherries." *Journal of Agricultural and Food Chemistry* 43 (2): 343–46. https://doi.org/10.1021/jf00050a015.

Gao, Yuan, Meng Wang, Nan Jiang, Yao Wang, and Xiaoyuan Feng. 2019. "Use of Ultra-Performance Liquid Chromatography-Tandem Mass Spectrometry on Sweet Cherries to Determine Phenolic Compounds in Peel and Flesh." *Journal of the Science of Food and Agriculture* 99 (7): 3555–62. https://doi.org/10.1002/jsfa.9576.

Garrido, M., D. Gonzalez-Gomez, M. Lozano, C. Barriga, S. D. Paredes, and A B Rodriguez. 2013. "Characterization and Trials of a Jerte Valley Cherry Product as a Natural Antioxidant-Enriched Supplement." *Italian Journal of Food Science* 25 (1): 90.

Garrido, María, Javier Espino, David González-Gómez, Mercedes Lozano, Carmen Barriga, Sergio D. Paredes, and Ana B. Rodríguez. 2012. "The Consumption of a Jerte Valley Cherry Product in Humans Enhances Mood, and Increases 5-Hydroxyindoleacetic Acid but Reduces Cortisol Levels in Urine." *Experimental Gerontology* 47 (8): 573–80. https://doi.org/10.1016/J.EXGER.2012.05.003.

Garrido, María, Javier Espino, David González-Gómez, Mercedes Lozano, Javier Cubero, Antonio F Toribio-Delgado, Juan I Maynar-Mariño, M Pilar Terrón, Juan L Muñoz, and José A Pariente. 2009. "A Nutraceutical Product Based on Jerte Valley Cherries Improves Sleep and Augments the Antioxidant Status in Humans." *E-SPEN, the European e-Journal of Clinical Nutrition and Metabolism* 4 (6): e321–23.

Ghosh, Dilip, and Tetsuya Konishi. 2007. "Anthocyanins and Anthocyanin-Rich Extracts: Role in Diabetes and Eye Function." *Asia Pac J Clin Nutr* 16 (2): 200–208. https://doi.org/10.3390/ijms13022472.

Gonçalves, Ana C., Márcio Rodrigues, Adriana O. Santos, Gilberto Alves, and Luís R. Silva. 2018. "Antioxidant Status, Antidiabetic Properties and Effects on Caco-2 Cells of Colored and Non-Colored Enriched Extracts of Sweet Cherry Fruits." *Nutrients* 10 (11). https://doi.org/10.3390/nu10111688.

Gonçalves, Berta, Anne Katrine Landbo, David Knudsen, Ana P. Silva, José Moutinho-Pereira, Eduardo Rosa, and Anne S. Meyer. 2004. "Effect of Ripeness and Postharvest Storage on the Phenolic Profiles of Cherries (*Prunus avium* L.)." *Journal of Agricultural and Food Chemistry* 52 (3): 523–30. https://doi.org/10.1021/jf030595s.

Gonçalves, Joana, Rodrigo Ramos, Ângelo Luís, Sandra Rocha, Tiago Rosado, Eugenia Gallardo, and Ana Paula Duarte. 2019. "Assessment of the Bioaccessibility and Bioavailability of the Phenolic Compounds of *Prunus avium* L. by in Vitro Digestion and Cell Model." *ACS Omega* 4: 7605–13. https://doi.org/10.1021/acsomega.8b03499.

Gonçalves, Joana, Rodrigo Ramos, Tiago Rosado, Eugenia Gallardo, and Ana Paula Duarte. 2019. "Development and Validation of a HPLC–

DAD Method for Quantification of Phenolic Compounds in Different Sweet Cherry Cultivars." *SN Applied Sciences* 1: 954. https://doi.org/10.1007/s42452-019-0680-4.

González-Gómez, D., M. Lozano, M. F. Fernández-León, M. C. Ayuso, M. J. Bernalte, and A. B. Rodríguez. 2009. "Detection and Quantification of Melatonin and Serotonin in Eight Sweet Cherry Cultivars (*Prunus avium* L.)." *European Food Research and Technology* 229 (2): 223–29. https://doi.org/10.1007/s00217-009-1042-z.

Grigoras, Cristina G., Emilie Destandau, Sandrine Zubrzycki, and Claire Elfakir. 2012. "Sweet Cherries Anthocyanins: An Environmental Friendly Extraction and Purification Method." *Separation and Purification Technology* 100 (October): 51–58. https://doi.org/10.1016/j.seppur.2012.08.032.

Habib, Muzammil, Mudassir Bhat, B. N. Dar, and Ali Abas Wani. 2017. "Sweet Cherries from Farm to Table: A Review." *Critical Reviews in Food Science and Nutrition* 57 (8): 1638–49. https://doi.org/10.1080/10408398.2015.1005831.

Haminiuk, Charles W. I., Giselle M. Maciel, Manuel S. V. Plata-Oviedo, and Rosane M. Peralta. 2012. "Phenolic Compounds in Fruits - an Overview." *International Journal of Food Science and Technology*. https://doi.org/10.1111/j.1365 2621.2012.03067.x.

Hanbali, Lama B, Rana M Ghadieh, Hiba A Hasan, Yasmine K Nakhal, and John J Haddad. 2013. "Measurement of Antioxidant Activity and Antioxidant Compounds Under Versatile Extraction Conditions: I. The Immuno-Biochemical Antioxidant Properties of Sweet Cherry (*Prunus avium*) Extracts." *Anti-Inflammatory & Anti-Allergy Agents in Medicinal Chemistry* 12 (2): 173–87. https://doi.org/10.2174/1871523011312020009.

He, YH, J Zhou, YS Wang, C Xiao, Y Tong, JCO Tang, ASC Chan, and AP Lu. 2006. "Anti-Inflammatory and Anti-Oxidative Effects of Cherries on Freund's Adjuvant-Induced Arthritis in Rats." *Scandinavian Journal of Rheumatology* 35 (5): 356–58. https://doi.org/10.1080/03009740600704155.

Homoki, Judit R., Andrea Nemes, Erika Fazekas, Gyöngyi Gyémánt, Péter Balogh, Ferenc Gál, Jamil Al-Asri, et al. 2016. "Anthocyanin Composition, Antioxidant Efficiency, and α-Amylase Inhibitor Activity of Different Hungarian Sour Cherry Varieties (*Prunus cerasus* L.)." *Food Chemistry* 194: 222–29. https://doi.org/10.1016/j.foodchem.2015.07.130.

Hou, De-Xing, Takashi Yanagita, Takuhiro Uto, Satoko Masuzaki, and Makoto Fujii. 2005. "Anthocyanidins Inhibit Cyclooxygenase-2 Expression in LPS-Evoked Macrophages: Structure-Activity Relationship and Molecular Mechanisms Involved." *Biochemical Pharmacology* 70 (3): 417–25. https://doi.org/10.1016/j.bcp.2005.05.003.

Ieri, Francesca, Patrizia Pinelli, and Annalisa Romani. 2012. "Simultaneous Determination of Anthocyanins, Coumarins and Phenolic Acids in Fruits, Kernels and Liqueur of *Prunus mahaleb* L." *Food Chemistry* 135 (4): 2157–62. https://doi.org/10.1016/j.foodchem.2012.07.083.

Ignat, Ioana, Irina Volf, and Valentin I Popa. 2013. "Analytical Methods of Phenolic Compounds." In *Natural Products: Phytochemistry, Botany and Metabolism of Alkaloids, Phenolics and Terpenes*, edited by Kishan Gopal Ramawat and Jean-Michel Mérillon, 2061–92. Berlin, Heidelberg: Springer Berlin Heidelberg. https://doi.org/10.1007/978-3-642-22144-6_56.

Jacob, Robert A., Giovanna M. Spinozzi, Vicky A. Simon, Darshan S. Kelley, Ronald L. Prior, Betty Hess-Pierce, and Adel A. Kader. 2003. "Consumption of Cherries Lowers Plasma Urate in Healthy Women." *The Journal of Nutrition* 133 (6): 1826–29. https://doi.org/10.1093/jn/133.6.1826.

Jayaprakasam, Bolleddula, L. Karl Olson, Robert E. Schutzki, Mei Hui Tai, and Muraleedharan G. Nair. 2006. "Amelioration of Obesity and Glucose Intolerance in High-Fat-Fed C57BL/6 Mice by Anthocyanins and Ursolic Acid in Cornelian Cherry (*Cornus mas*)." *Journal of Agricultural and Food Chemistry* 54 (1): 243–48. https://doi.org/10.1021/jf0520342.

Jayaprakasam, Bolleddula, Shaiju K Vareed, L Karl Olson, and Muraleedharan G Nair. 2005. "Insulin Secretion by Bioactive Anthocyanins and Anthocyanidins Present in Fruits." *Journal of Agricultural and Food Chemistry* 53 (1): 28–31.

Kappel Frank, Fisher-Fleming Bob and Hogue Eugene. 1996. "*Fruit Characteristics and Sensory Attributes of an Ideal Sweet Cherry*" 31 (3): 443–46.

Kelebek, Hasim, and Serkan Selli. 2011. "Evaluation of Chemical Constituents and Antioxidant Activity of Sweet Cherry (*Prunus avium* L.) Cultivars." *International Journal of Food Science & Technology* 46 (12): 2530–37. https://doi.org/10.1111/j.1365-2621.2011.02777.x.

Kelley, Darshan S., Yuriko Adkins, and Kevin D. Laugero. 2018. "A Review of the Health Benefits of Cherries." *Nutrients* 10 (3): 368. https://doi.org/10.3390/nu10030368.

Kelley, Darshan S., Yuriko Adkins, Aurosis Reddy, Leslie R. Woodhouse, Bruce E. Mackey, and Kent L. Erickson. 2013. "Sweet Bing Cherries Lower Circulating Concentrations of Markers for Chronic Inflammatory Diseases in Healthy Humans." *The Journal of Nutrition* 143 (3): 340–44. https://doi.org/10.3945/jn.112.171371.

Kelley, Darshan S., Reuven Rasooly, Robert A. Jacob, Adel A. Kader, and Bruce E. Mackey. 2006. "Consumption of Bing Sweet Cherries Lowers Circulating Concentrations of Inflammation Markers in Healthy Men and Women." *The Journal of Nutrition* 136 (4): 981–86. https://doi.org/10.1093/jn/136.4.981.

Kent, Katherine, Karen Charlton, Steven Roodenrys, Marijka Batterham, Jan Potter, Victoria Traynor, Hayley Gilbert, Olivia Morgan, and Rachelle Richards. 2017. "Consumption of Anthocyanin-Rich Cherry Juice for 12 Weeks Improves Memory and Cognition in Older Adults with Mild-to-Moderate Dementia." *European Journal of Nutrition* 56 (1): 333–41. https://doi.org/10.1007/s00394-015-1083-y.

Kim, Dae-Ok, Ho Jin Heo, Young Jun Kim, Hyun Seuk Yang, and Chang Y. Lee. 2005. "Sweet and Sour Cherry Phenolics and Their Protective Effects on Neuronal Cells." *Journal of Agricultural and Food Chemistry* 53 (26): 9921–27. https://doi.org/10.1021/jf0518599.

Krinsky, Norman I., and Kyung-Jin Yeum. 2003. "Carotenoid–radical Interactions." *Biochemical and Biophysical Research Communications* 305 (3): 754–60. https://doi.org/10.1016/S0006-291X(03)00816-7.

Kulisic-Bilusic, Tea, Kerstin Schnäbele, Ingrid Schmöller, Verica Dragovic-Uzelac, Anita Krisko, Branka Dejanovic, Mladen Milos, and Greta Pifat. 2009. "Antioxidant Activity versus Cytotoxic and Nuclear Factor Kappa B Regulatory Activities on HT-29 Cells by Natural Fruit Juices." *European Food Research and Technology* 228 (3): 417–24. https://doi.org/10.1007/s00217-008-0948-1.

Lachin, Tahsini. 2014. "Effect of Antioxidant Extract from Cherries on Diabetes." *Recent Patents on Endocrine, Metabolic & Immune Drug Discovery* 8 (1): 67–74. https://doi.org/10.2174/1872214808666140121151334.

Lachin, Tahsini, and Heydari Reza. 2012. "Anti Diabetic Effect of Cherries in Alloxan Induced Diabetic Rats." *Recent Patents on Endocrine, Metabolic & Immune Drug Discovery* 6 (1): 67–72. https://doi.org/10.2174/187221412799015308.

Lazzè, Maria Claudia, Monica Savio, Roberto Pizzala, Ornella Cazzalini, Paola Perucca, Anna Ivana Scovassi, Lucia Anna Stivala, and Livia Bianchi. 2004. "Anthocyanins Induce Cell Cycle Perturbations and Apoptosis in Different Human Cell Lines." *Carcinogenesis* 25 (8): 1427–33. https://doi.org/10.1093/carcin/bgh138.

Li, An Na, Sha Li, Yu Jie Zhang, Xiang Rong Xu, Yu Ming Chen, and Hua Bin Li. 2014. "Resources and Biological Activities of Natural Polyphenols." *Nutrients*. MDPI AG. https://doi.org/10.3390/nu6126020.

Lim, T. K. 2012. "*Prunus avium*." In *Edible Medicinal And Non-Medicinal Plants*, 451–62. Dordrecht: Springer Netherlands. https://doi.org/10.1007/978-94-007-1764-0_1.

Lynn, Anthony, Shilpa Mathew, Chris T Moore, Jean Russell, Emma Robinson, Vithleem Soumpasi, and Margo E Barker. 2014. "Effect of a Tart Cherry Juice Supplement on Arterial Stiffness and Inflammation in Healthy Adults: A Randomised Controlled Trial." *Plant Foods for*

Human Nutrition 69 (2): 122–27. https://doi.org/10.1007/s11130-014-0409-x.

Martin, Keith R, Jennie Bopp, Lacey Burrell, and Ginger Hook. 2011. "*The Effect of 100% Tart Cherry Juice on Serum Uric Acid Levels, Biomarkers of Inflammation and Cardiovascular Disease Risk Factors.*" Federation of American Societies for Experimental Biology.

Martini, Serena, Angela Conte, and Davide Tagliazucchi. 2017. "Phenolic Compounds Profile and Antioxidant Properties of Six Sweet Cherry (*Prunus avium*) Cultivars." *Food Research International* 97: 15–26. https://doi.org/10.1016/J.FOODRES.2017.03.030.

Mazza, G., L. Fukumoto, P. Delaquis, B. Girard, and B. Ewert. 1999. "Anthocyanins, Phenolics, and Color of Cabernet Franc, Merlot, and Pinot Noir Wines from British Columbia." *Journal of Agricultural and Food Chemistry* 47 (10): 4009–17. https://doi.org/10.1021/jf990449f.

McCune, Letitia M., Chieri Kubota, Nicole R. Stendell-Hollis, and Cynthia A. Thomson. 2011. "Cherries and Health: A Review." *Critical Reviews in Food Science and Nutrition*. https://doi.org/10.1080/10408390903001719.

Meiers, Susanne, Monika Kemény, Ulrike Weyand, Robert Gastpar, Erwin Von Angerer, and Doris Marko. 2001. "The Anthocyanidins Cyanidin and Delphinidin Are Potent Inhibitors of the Epidermal Growth-Factor Receptor." *Journal of Agricultural and Food Chemistry* 49 (2): 958–62. https://doi.org/10.1021/jf0009100.

Mikulic-Petkovsek, Maja, Franci Stampar, Robert Veberic, and Helena Sircelj. 2016. "Wild Prunus Fruit Species as a Rich Source of Bioactive Compounds." *Journal of Food Science* 81 (8): C1928–37. https://doi.org/10.1111/1750-3841.13398.

Minekus, M., M. Alminger, P. Alvito, S. Ballance, T. Bohn, C. Bourlieu, F. Carrière, et al. 2014. "A Standardised Static in Vitro Digestion Method Suitable for Food-an International Consensus." *Food and Function* 5 (6): 1113–24. https://doi.org/10.1039/c3fo60702j.

Mulabagal, Vanisree, Gregory A. Lang, David L. Dewitt, Sanjeev S. Dalavoy, and Muraleedharan G. Nair. 2009. "Anthocyanin Content, Lipid Peroxidation and Cyclooxygenase Enzyme Inhibitory Activities of Sweet and Sour Cherries." *Journal of Agricultural and Food Chemistry* 57 (4): 1239–46. https://doi.org/10.1021/jf8032039.

Mülleder, Ursula, Michael Murkovic, and Werner Pfannhauser. 2002. "Urinary Excretion of Cyanidin Glycosides." *Journal of Biochemical and Biophysical Methods* 53 (1–3): 61–66. https://doi.org/10.1016/s0165-022x(02)00093-3.

Nawirska-Olszańska, Agnieszka, Joanna Kolniak-Ostek, Maciej Oziembłowski, Alena Ticha, Radomir Hyšpler, Zdenek Zadak, Pavla Židová, and Frantisek Paprstein. 2017. "Comparison of Old Cherry Cultivars Grown in Czech Republic by Chemical Composition and Bioactive Compounds." *Food Chemistry* 228 (August): 136–42. https://doi.org/10.1016/j.foodchem.2017.01.154.

Nowicka, Paulina, Aneta Wojdyło, and Justyna Samoticha. 2016. "Evaluation of Phytochemicals, Antioxidant Capacity, and Antidiabetic Activity of Novel Smoothies from Selected Prunus Fruits." *Journal of Functional Foods* 25: 397–407. https://doi.org/10.1016/j.jff.2016.06.024.

Pacifico, Severina, Antimo Di Maro, Milena Petriccione, Silvia Galasso, Simona Piccolella, Antonella M A Di Giuseppe, Marco Scortichini, and Pietro Monaco. 2014. "Chemical Composition, Nutritional Value and Antioxidant Properties of Autochthonous *Prunus avium* Cultivars from Campania Region." *Food Research International (Ottawa, Ont.)* 64 (October): 188–99. https://doi.org/10.1016/j.foodres.2014.06.020.

Paganga, G, and C A Rice-Evans. 1997. "The Identification of Flavonoids as Glycosides in Human Plasma." *FEBS Letters* 401 (1): 78–82. https://doi.org/10.1016/s0014-5793(96)01442-1.

Papapetros, Spyridon, Artemis Louppis, Ioanna Kosma, Stavros Kontakos, Anastasia Badeka, and Michael G. Kontominas. 2018. "Characterization and Differentiation of Botanical and Geographical Origin of Selected Popular Sweet Cherry Cultivars Grown in Greece." *Journal of Food Composition and Analysis* 72: 48–56. https://doi.org/10.1016/J.JFCA.2018.06.006.

Papapetros, Spyridon, Artemis Louppis, Ioanna Kosma, Stavros Kontakos, Anastasia Badeka, Chara Papastephanou, and Michael G. Kontominas.

2019. "Physicochemical, Spectroscopic and Chromatographic Analyses in Combination with Chemometrics for the Discrimination of Four Sweet Cherry Cultivars Grown in Northern Greece." *Foods* 8 (10): 442–56. https://doi.org/10.3390/foods8100442.

Papp, Nóra, Blanka Szilvássy, László Abrankó, Tibor Szabó, Péter Pfeiffer, Zoltán Szabó, József Nyéki, Sezai Ercisli, Éva Stefanovits-Bányai, and Attila Hegedűs. 2010. "Main Quality Attributes and Antioxidants in Hungarian Sour Cherries: Identification of Genotypes with Enhanced Functional Properties." *International Journal of Food Science & Technology* 45 (2): 395–402. https://doi.org/10.1111/j.1365-2621.2009.02168.x.

Piccolella, Simona, Giuseppina Crescente, Paola Nocera, Francesca Pacifico, Lorenzo Manti, and Severina Pacifico. 2018. "Ultrasound-Assisted Aqueous Extraction, LC-MS/MS Analysis and Radiomodulating Capability of Autochthonous Italian Sweet Cherry Fruits." *Food & Function* 9 (3): 1840–49. https://doi.org/10.1039/c7fo01977g.

Platt, K. L., R. Edenharder, S. Aderhold, E. Muckel, and H. Glatt. 2010. "Fruits and Vegetables Protect against the Genotoxicity of Heterocyclic Aromatic Amines Activated by Human Xenobiotic-Metabolizing Enzymes Expressed in Immortal Mammalian Cells." *Mutation Research - Genetic Toxicology and Environmental Mutagenesis* 703 (2): 90–98. https://doi.org/10.1016/j.mrgentox.2010.08.007.

Redondo, Diego, Esther Arias, Rosa Oria, and María E. Venturini. 2017. "Thinned Stone Fruits Are a Source of Polyphenols and Antioxidant Compounds." *Journal of the Science of Food and Agriculture* 97 (3): 902–10. https://doi.org/10.1002/jsfa.7813.

Rio, D. Del, L. G. Costa, M. E. J. Lean, and A. Crozier. 2010. "Polyphenols and Health: What Compounds Are Involved?" *Nutrition, Metabolism and Cardiovascular Diseases*. https://doi.org/10.1016/j.numecd.2009.05.015.

Rio, Daniele Del, Ana Rodriguez-Mateos, Jeremy P. E. Spencer, Massimiliano Tognolini, Gina Borges, and Alan Crozier. 2013.

"Dietary (Poly)Phenolics in Human Health: Structures, Bioavailability, and Evidence of Protective Effects against Chronic Diseases." *Antioxidants and Redox Signaling*. https://doi.org/10.1089/ars.2012.4581.

Rosado, T., I. Henriques, E. Gallardo, and A. P. Duarte. 2017. "Determination of Melatonin Levels in Different Cherry Cultivars by High-Performance Liquid Chromatography Coupled to Electrochemical Detection." *European Food Research and Technology*. https://doi.org/10.1007/s00217-017-2880-8.

Rosado, T., M. Barroso, A. P. Duarte, and E. Gallardo. 2017. "Determination of Melatonin in Biological and Non-Biological Samples." In *Melatonin: Medical Uses and Role in Health and Disease*, edited by Lore Correia and Germaine Mayers, 1–68. New York: Nova Science Publisher Inc. https://novapublishers.com/shop/melatonin-medical-uses-and-role-in-health-and-disease/.

Saleh, Fatima A., Nada El-Darra, and Karim Raafat. 2019. *Red Sour Cherry for the Treatment of Diabetes Mellitus*. *Bioactive Food as Dietary Interventions for Diabetes*. 2nd ed. Elsevier Inc. https://doi.org/10.1016/b978-0-12-813822-9.00033-3.

Sancho, Renata A.Soriano, and Glaucia Maria Pastore. 2012. "Evaluation of the Effects of Anthocyanins in Type 2 Diabetes." *Food Research International* 46 (1): 378–86. https://doi.org/10.1016/j.foodres.2011.11.021.

Scalbert, Augustin, Claudine Manach, Christine Morand, Christian Rémésy, and Liliana Jiménez. 2005. "Dietary Polyphenols and the Prevention of Diseases." *Critical Reviews in Food Science and Nutrition*. https://doi.org/10.1080/1040869059096.

Seeram, N. P., R. A. Momin, M. G. Nair, and L. D. Bourquin. 2001. "Cyclooxygenase Inhibitory and Antioxidant Cyanidin Glycosides in Cherries and Berries." *Phytomedicine* 8 (5): 362–69. https://doi.org/10.1078/0944-7113-00053.

Seeram, Navindra P, Yanjun Zhang, and Muraleedharan G Nair. 2003. "Inhibition of Proliferation of Human Cancer Cells and Cyclooxygenase Enzymes by Anthocyanidins and Catechins."

Nutrition and Cancer 46 (1): 101–6. https://doi.org/10.1207/S15327914NC4601_13.

Serra, Ana Teresa, Rui O. Duarte, Maria R. Bronze, and Catarina M.M. Duarte. 2011. "Identification of Bioactive Response in Traditional Cherries from Portugal." *Food Chemistry* 125 (2): 318–25. https://doi.org/10.1016/j.foodchem.2010.07.088.

Serra, Ana Teresa, Inês J. Seabra, Mara E. M. Braga, M. R. Bronze, Hermínio C. De Sousa, and Catarina M. M. Duarte. 2010. "Processing Cherries (*Prunus avium*) Using Supercritical Fluid Technology. Part 1: Recovery of Extract Fractions Rich in Bioactive Compounds." *Journal of Supercritical Fluids* 55 (1): 184–91. https://doi.org/10.1016/j.supflu.2010.06.005.

Serradilla, Manuel Joaquín, Alejandro Hernández, Margarita López-Corrales, Santiago Ruiz-Moyano, María de Guía Córdoba, and Alberto Martín. 2015. "Composition of the Cherry (*Prunus avium* L. and Prunus Cerasus L.; Rosaceae)." In *Nutritional Composition of Fruit Cultivars*, 127–47. Elsevier Inc. https://doi.org/10.1016/B978-0-12-408117-8.00006-4.

Serradilla, Manuel Joaquín, Mercedes Lozano, María Josefa Bernalte, María Concepción Ayuso, Margarita López-Corrales, and David González-Gómez. 2011. "Physicochemical and Bioactive Properties Evolution during Ripening of 'Ambrunés' Sweet Cherry Cultivar." *LWT - Food Science and Technology* 44 (1): 199–205. https://doi.org/10.1016/j.lwt.2010.05.036.

Serrano, María, Fabián Guillén, Domingo Martínez-Romero, Salvador Castillo, and Daniel Valero. 2005. "Chemical Constituents and Antioxidant Activity of Sweet Cherry at Different Ripening Stages." *Journal of Agricultural and Food Chemistry* 53 (7): 2741–45. https://doi.org/10.1021/jf0479160.

Seymour, E. M., Sarah K. Lewis, Daniel E. Urcuyo-Llanes, Ignasia I. Tanone, Ara Kirakosyan, Peter B. Kaufman, and Steven F. Bolling. 2009. "Regular Tart Cherry Intake Alters Abdominal Adiposity, Adipose Gene Transcription, and Inflammation in Obesity-Prone Rats

Fed a High Fat Diet." *Journal of Medicinal Food* 12 (5): 935–42. https://doi.org/10.1089/jmf.2008.0270.

Seymour, E. Mitchell, Andrew A. M. Singer, Ara Kirakosyan, Daniel E. Urcuyo-Llanes, Peter B Kaufman, and Steven F Bolling. 2008. "Altered Hyperlipidemia, Hepatic Steatosis, and Hepatic Peroxisome Proliferator-Activated Receptors in Rats with Intake of Tart Cherry." *Journal of Medicinal Food* 11 (2): 252–59. https://doi.org/10.1089/jmf.2007.658.

Shih, Ping-Hsiao, Chi-Tai Yeh, and Gow-Chin Yen. 2005. "Effects of Anthocyanidin on the Inhibition of Proliferation and Induction of Apoptosis in Human Gastric Adenocarcinoma Cells." *Food and Chemical Toxicology : An International Journal Published for the British Industrial Biological Research Association* 43 (10): 1557–66. https://doi.org/10.1016/j.fct.2005.05.001.

Silva, Gonçalo R., Cátia V. Vaz, Beatriz Catalão, Susana Ferreira, Henrique J. Cardoso, Ana Paula Duarte, and Sílvia Socorro. 2019. "Sweet Cherry Extract Targets the Hallmarks of Cancer in Prostate Cells: Diminished Viability, Increased Apoptosis and Suppressed Glycolytic Metabolism." *Nutrition and Cancer*, September, 1–15. https://doi.org/10.1080/01635581.2019.1661502.

Singleton, V. L., and Joseph A. Rossi. 1965. "Colorimetry of Total Phenolics with Phosphomolybdic-Phosphotungstic Acid Reagents." *American Journal of Enology and Viticulture* 16 (3).

Snyder, Sarah M, Bingxin Zhao, Ting Luo, Clive Kaiser, George Cavender, Jill Hamilton-Reeves, Debra K Sullivan, and Neil F Shay. 2016. "Consumption of Quercetin and Quercetin-Containing Apple and Cherry Extracts Affects Blood Glucose Concentration, Hepatic Metabolism, and Gene Expression Patterns in Obese C57BL/6J High Fat–Fed Mice." *The Journal of Nutrition* 146 (5): 1001–7. https://doi.org/10.3945/jn.115.228817.

Subash, Selvaraju, Musthafa Mohamed Essa, Samir Al-Adawi, Mushtaq A. Memon, Thamilarasan Manivasagam, and Mohammed Akbar. 2014. "Neuroprotective Effects of Berry Fruits on Neurodegenerative

Diseases." *Neural Regeneration Research* 9 (16): 1557–66. https://doi.org/10.4103/1673-5374.139483.

United States Department of Agriculture. 2019. "*FoodData Central.*" 2019. https://fdc.nal.usda.gov/fdc-app.html#/?query=sweet cherries.

Usenik, Valentina, Jerneja Fabčič, and Franci Štampar. 2008. "Sugars, Organic Acids, Phenolic Composition and Antioxidant Activity of Sweet Cherry (*Prunus avium* L.)." *Food Chemistry* 107 (1): 185–92. https://doi.org/10.1016/j.foodchem.2007.08.004.

Uttara, Bayani, Ajay Singh, Paolo Zamboni, and R. Mahajan. 2009. "Oxidative Stress and Neurodegenerative Diseases: A Review of Upstream and Downstream Antioxidant Therapeutic Options." *Current Neuropharmacology* 7 (1): 65–74. https://doi.org/10.2174/1570159 09787602823.

Valero, Daniel, Huertas M. Díaz-Mula, Pedro Javier Zapata, Salvador Castillo, Fabián Guillén, Domingo Martínez-Romero, and María Serrano. 2011. "Postharvest Treatments with Salicylic Acid, Acetylsalicylic Acid or Oxalic Acid Delayed Ripening and Enhanced Bioactive Compounds and Antioxidant Capacity in Sweet Cherry." *Journal of Agricultural and Food Chemistry* 59 (10): 5483–89. https://doi.org/10.1021/jf200873j.

Vavoura, Maria V., Anastasia V. Badeka, Stavros Kontakos, and Michael G. Kontominas. 2015. "Characterization of Four Popular Sweet Cherry Cultivars Grown in Greece by Volatile Compound and Physicochemical Data Analysis and Sensory Evaluation." *Molecules* 20 (2): 1922–40. https://doi.org/10.3390/molecules20021922.

Vermerris, Wilfred, and Ralph Nicholson. 2006. *Phenolic Compound Biochemistry. Phenolic Compound Biochemistry.* Springer Netherlands. https://doi.org/10.1007/978-1-4020-5164-7.

Versantvoort, Carolien H. M., Agnes G. Oomen, Erwin Van De Kamp, Cathy J. M. Rompelberg, and Adriënne J. A. M. Sips. 2005. "Applicability of an in Vitro Digestion Model in Assessing the Bioaccessibility of Mycotoxins from Food." *Food and Chemical Toxicology* 43 (1): 31–40. https://doi.org/10.1016/j.fct.2004.08.007.

Vinitha, Edula, Hanish J. C. Singh, Rahul Motiram Kakalij, Rahul Padmakar Kshirsagar, Boyina Hemanth Kumar, and Prakash V. Diwan. 2014. "Neuroprotective Effect of Prunus Avium on Streptozotocin Induced Neurotoxicity in Mice." *Biomedicine & Preventive Nutrition* 4 (4): 519–25. https://doi.org/10.1016/J.BIONUT.2014.08.004.

Wang, J., and G. Mazza. 2002. "Inhibitory Effects of Anthocyanins and Other Phenolic Compounds on Nitric Oxide Production in LPS/IFN-γ-Activated RAW 264.7 Macrophages." *Journal of Agricultural and Food Chemistry* 50 (4): 850–57. https://doi.org/10.1021/jf010976a.

Wang, Jian, and G. Mazza. 2002. "Effects of Anthocyanins and Other Phenolic Compounds on the Production of Tumor Necrosis Factor α in LPS/IFN-γ-Activated RAW 264.7 Macrophages." *Journal of Agricultural and Food Chemistry* 50 (15): 4183–89. https://doi.org/10.1021/jf011613d.

Wang, Li Shu, and Gary D. Stoner. 2008. "Anthocyanins and Their Role in Cancer Prevention." *Cancer Letters*. https://doi.org/10.1016/j.canlet.2008.05.020.

Wang, Meng, Nan Jiang, Yao Wang, Dongmei Jiang, and Xiaoyuan Feng. 2017. "Characterization of Phenolic Compounds from Early and Late Ripening Sweet Cherries and Their Antioxidant and Antifungal Activities." *Journal of Agricultural and Food Chemistry* 65 (26): 5413–20. https://doi.org/10.1021/acs.jafc.7b01409.

Wen, Ya Qin, Fei He, Bao Qing Zhu, Yi Bin Lan, Qiu Hong Pan, Chun You Li, Malcolm J. Reeves, and Jun Wang. 2014. "Free and Glycosidically Bound Aroma Compounds in Cherry (*Prunus avium* L.)." *Food Chemistry* 152 (June): 29–36. https://doi.org/10.1016/j.foodchem.2013.11.092.

Wenzel, Amy, Erin N. Haugen, Lydia C. Jackson, and Jennifer R. Brendle. 2005. "Anxiety Symptoms and Disorders at Eight Weeks Postpartum." *Journal of Anxiety Disorders* 19 (3): 295–311. https://doi.org/10.1016/J.JANXDIS.2004.04.001.

Wu, Tao, Jinjin Yin, Guohua Zhang, Hairong Long, and Xiaodong Zheng. 2016. "Mulberry and Cherry Anthocyanin Consumption Prevents

Oxidative Stress and Inflammation in Diet-Induced Obese Mice." *Molecular Nutrition and Food Research* 60 (3): 687–94. https://doi.org/10.1002/mnfr.201500734.

Zhang, Hua, and Rong Tsao. 2016. "Dietary Polyphenols, Oxidative Stress and Antioxidant and Anti-Inflammatory Effects." *Current Opinion in Food Science*. Elsevier Ltd. https://doi.org/10.1016/j.cofs.2016.02.002.

Zhang, Yuqing, Tuhina Neogi, Clara Chen, Christine Chaisson, David J Hunter, and Hyon K Choi. 2012. "Cherry Consumption and Decreased Risk of Recurrent Gout Attacks." *Arthritis and Rheumatism* 64 (12): 4004–11. https://doi.org/10.1002/art.34677.

Zhang, Zhaoyan, Hong Zheng, Kun Liang, Hui Wang, Sumei Kong, Jinna Hu, Fang Wu, and Gang Sun. 2015. "Functional Degeneration in Dorsal and Ventral Attention Systems in Amnestic Mild Cognitive Impairment and Alzheimer's Disease: An FMRI Study." *Neuroscience Letters* 585 (January): 160–65. https://doi.org/10.1016/J.NEULET.2014.11.050.

Zhao, Handong, Bangdi Liu, Wanli Zhang, Jiankang Cao, and Weibo Jiang. 2019. "Enhancement of Quality and Antioxidant Metabolism of Sweet Cherry Fruit by Near-Freezing Temperature Storage." *Postharvest Biology and Technology* 147 (January): 113–22. https://doi.org/10.1016/J.POSTHARVBIO.2018.09.013.

In: *Prunus*
Editor: William Schneider
ISBN: 978-1-53617-755-8

Chapter 3

ECOPHYSIOLOGICAL MECHANISMS OF THE INVASION OF BLACK CHERRY IN EUROPEAN FORESTS

*Piotr Robakowski**
Poznań University of Life Sciences, Faculty of Forestry,
Department of Silviculture,
Chair of Ecological Foundation of Silviculture,
Wojska Polskiego, Poznań, Poland

ABSTRACT

Black cherry (*Prunus serotina* Ehrh.), native to North America, is a tree species appreciated for its valuable wood and ornamental leaves and flowers. Black cherry has been cultivated as an ornamental tree in European gardens, and in Central Europe it has often been planted as a biocenotic admixture under a canopy of Scots pine. In Europe, outside its native North American range, it becomes naturalized and is a highly invasive neophyte.

* Corresponding Author's Email: piotr.robakowski@up.poznan.pl.

When the competitiveness and invasiveness of black cherry are compared between its native and exotic range, the duality of the species' behavior can be observed. Its high level of invasiveness in the non-native range can be explained by the 'enemy release' or 'new weapon' hypotheses, as can species-specific growth, reproduction and physiological attributes. In the European occurrence range, growth dynamics of black cherry at a young age are greater than those of co-occurring native tree species. Seedlings allocate more biomass to above-ground organs than to roots when compared with native competitors. Black cherry grows slowly in strong shade and responds to increasing irradiance with faster growth rates. Its large biomass and leaf area enhance light capture and net CO_2 assimilation rates, giving black cherry an advantage over competitors. Both shade and sun leaves of black cherry have shown great plasticity regarding leaf-absorbed energy partitioning. These leaves are rich in nitrogen and may improve soil fertility, but they also contain the toxic cyanogenic glycosides prunasin and amygdalin. These allelochemicals, also found in the seeds and roots of this species, effectively play a protective role against herbivores, but may also inhibit growth of native plants occurring in the same niches. In its non-native range, black cherry is not browsed by ungulates and hosts few pests that are able to feed on tissues containing cyanogenic glycosides. This tree reproduces vegetatively by sprouts and generatively by seeds, which are often dispersed over long distances by birds.

At the seedling and sapling stages, all these traits allow black cherry to outcompete most European tree species. Its strategy of higher biomass investment in above-ground organs, coupled with an increase in light capture and faster net CO_2 assimilation rates, is more important for the success of its invasion outside its native range than any potential allelopathic effects on neighboring native plant species. In European forests, the invasive black cherry has been intensively fought off using mechanical and chemical methods, but without success. The enlargement of its distribution through its invasion of European forests suggests that it may adapt to global climate changes faster than do native species.

Keywords: allelopathy, growth, life strategy, photosynthesis, *Prunus serotina*

1. Taxonomy and Characteristics

Black cherry belongs to the family *Rosaceae*. The preferred scientific name for this species is *Prunus serotina* (Ehrh.), but other names have been used: *Padus serotina* (Borkh.), *Prunus virginiana* (L.), *Padus virginiana* (L) Mill. and *Padus serotina* (Ehrh.) J. Agardh. In North America, the species has been also known by the common names wild black cherry, rum cherry and mountain black cherry (Marquis 1990). In its natural occurrence range, black cherry comprises a complex of five morphologically different subspecies growing in significantly different niches. The subspecies *eximia*, *hirsuta* and *serotina* are distributed in more humid and cold environments, whereas subsp. *virens* prefers drier and warmer environments. The subspecies *capuli* exhibits the greatest environmental heterogeneity of the five (Guzmán et al. 2018; Segura et al. 2018). Although the subspecies show some morphological and ecological variation, they have not been recognized in the European non-native occurrence range of *P. serotina*. Instead, the following varieties have been distinguished: *Prunus serotina* var. *alabamensis* (C. Mohr) Little and *Prunus serotina* var. *rufula* (Wooton & Standl.) McVaugh, both commonly known as southwestern chokecherry or black cherry; and *Prunus serotina* var. *salicifolia* (Kunth) Koehne and *Prunus serotina* var. *serotina* Ehrh., both known as black cherry (Robakowski and Stachnowicz 2000).

Black cherry is a deciduous, single-stemmed, medium- to large-sized tree. In mixed hardwood forest, it can also be a small, contorted, short-lived understory tree or shrub. The growth form of *P. serotina* depends on provenance, local soil and climate conditions, and anthropogenic activity. When black cherry is cut, it sprouts from the trunk and roots, growing as a shrub. It has been observed that the species occurs more frequently as a tree in its native range, and more frequently as a shrub in its introduced range. This difference between native and introduced plants may result from variation in precipitation conditions during growth. In the central European range of *P. serotina*, the mean annual sum of precipitation is approximately 600 mm. In contrast, on the Allegheny Plateau of Pennsylvania and New York, there is more precipitation, ranging from 970 to 1120 mm annually (Starfinger et al. 2003). In the eastern United States, black cherry typically reaches 38 m in height and 1.2 m or more in diameter at breast height. South-western varieties are much smaller (Marquis 1990). In Central Europe, black cherry is often a shrub or, more rarely, a tree up to 20 m in height often growing under the canopy of monospecific, even-aged Scots pine (*Pinus sylvestris* L.) stands (Starfinger et al. 2003; Halarewicz 2011). The species is important for wildlife due to its flowers and fruits. Flowering occurs before leaves appear, usually in May (Hough 1965; Marquis 1990). Black cherry flowers are white, solitary and borne in umbel-like racemes. The flowers are insect pollinated, with bees feeding on the nectar of the blossoms and providing highly appreciated honey (Marquis 1990; Robakowski and Stachnowicz 2000).

2. Geographical Distribution

Black cherry is native to North America. Its range comprises central and eastern parts of the USA, south-eastern parts of Canada, Mexico and parts of northern South America, down to Guatemala. Several varieties of *P. serotina* grow outside this area, however (Marquis 1990). Black cherry was one of the first North American trees to be cultivated as ornamental trees in European gardens because of its beautiful flowers and glossy leaves, which in autumn change colors from green through to yellow and reddish. The species was introduced into England in 1629 (Hough 1957), but has also been planted in other western and central European countries. In Belgium, the Netherlands and Germany, *P. serotina* was introduced for fire protection and soil amelioration in pine plantations on poor sandy soils (Starfinger et al. 2003; Verheyen et al. 2007). In Poland, *P. serotina* was on the list of species recommended to be planted as a component of undergrowth in dry, fresh coniferous and mixed fresh coniferous forests to ameliorate soil conditions and to increase biodiversity. The planting of *P. serotina* in central European forests was finally forbidden in the 1980s and, since that time seedlings of this species have not been produced in forest nurseries. The large-scale planting of *P. serotina* in forests and its high reproductive potential, however, have caused a rapid spread of the species under a canopy of Scots pine forests, in gaps and in woodlands. In many places, black cherry becomes naturalized and appears to be a highly invasive neophyte (Starfinger 1997; Möllerová 2005; Csiszár et al. 2013;

Bijak et al. 2014; Halarewicz and Żołnierz 2014). Presently, the species grows spontaneously in many European countries, from France to Poland and Romania, and from Denmark to Italy (Pairon et al. 2010).

3. Ecology

3.1. Reproduction

Black cherry can reproduce via seeds or vegetatively by sprouting from stumps or roots. In its native range, it begins to fruit at the age of 10 years (Marquis 1990). In contrast, in its introduced range, trees even as young as 5.2 years may blossom (Deckers et al. 2005). The dates of flowering and fruiting depend on the light environment during growth; in the open blossoming begins earlier, at the age of 5 – 6 years, whereas under the tree canopy, blossoming occurs at 20 years. In natural stands, 30- to 100-year-old trees produce the greatest number of seeds. In Pennsylvania and New York, black cherry flowers usually appear at nearly full leaf expansion, around May 15 to May 20. Flower development in other parts of the range varies with climate, from the end of March in Texas to the first week of June in Québec, Canada. On the Allegheny Plateau, fruit ripening and seed fall occur between August 15 and mid-September. In the south-eastern United States, fruits ripen in late June and seed fall is complete by early July (Marquis 1990). In Europe, black cherry flowers appear in late spring, later than the flowers of the European species *Prunus padus* (L.). Fruit, which is edible and nearly black in color, ripens in September or later. As a tree, black cherry can reach approximately 250 years of age (Starfinger 1997).

Seeds are dispersed by gravity and animals, mainly birds, and require cold stratification to germinate. Seeds that have passed through the digestive tracts of birds exhibit higher germination rates than undigested ones. *Prunus serotina* balances a fairly low germination rate and a high rate of seed predation with its ability to germinate under a wider range of physical conditions. Moreover, germination can be delayed depending on climate conditions (Auclair and Cottam 1971; van den Tweel and Eijsackers 1987). Seeds can stay in the soil for three to four years, waiting for the most favorable conditions for germination. More seeds germinate in sunlight than under a canopy of trees, and germination occurs more often in litter than in mineral soil (Smith 1975).

Black cherry also sprouts from stumps following cutting or fire. Sprouts often grow faster than seedlings during their first 20 – 30 years. The wood from the first generation of sprouts can be of high quality and is often used for sawtimber, but later generations of sprouts degenerate in quality, likely due to their use of an old and dysfunctional root system (Wendel 1975).

3.2. Light Requirements

As a shade-intolerant species, black cherry grows willingly in forest clearings and gaps, but seedlings can also be found beneath the dense canopy. Slow-growing seedlings and saplings can survive even under the deep shade of a pine thicket (Figures 1, 2). Under non-limiting growth conditions, potted *P. serotina* seedlings were able to grow in 5 or 10% of full sunlight, under shading net, but their seasonal height and stem diameter growth rates were low. At a young age, the species can be classified as moderately shade-tolerant, showing its highest growth rates in 100% sun, lower rates in 25% sun and nearly no growth in 10% sun (Robakowski et al. 2018). Characteristics of shade-acclimated seedlings differ remarkably from those of full light acclimated ones, with shade seedlings being approximately 5-fold smaller, with dark-green, large, thin leaves (Figure 3a, b).

In the forest, the age and height structure of the black cherry population is heterogeneous because under dense canopy shade, the species develops the 'sit and wait' life strategy (Closset-Kopp et al. 2007). Seedlings and saplings representing different generations wait in deep shade (lower than 10% of full sunlight) for an improvement in the light environment. When young cherries are acclimated to shade for several years and then abruptly exposed to high sunlight, their growth rates increase remarkably. Silvertown (1982) called this behavior 'Oskar syndrome' after the main character of the Günter Grass novel "Die Blechtrommel" (English: "The Tin Drum") who stopped growing at the age of three. 'Oskar syndrome' has been defined as a dynamic increase in the growth of a shade acclimated plant following exposure to a brighter light environment (Silvertown 1982; Closset-Kopp et al. 2007, 2011). This behavior has been observed under the canopy of Scots pine thickets, where black cherry seedlings grew for several years in deep shade and then were abruptly exposed to high levels of sunlight after thinning (personal observations). According to Halarewicz and Żołnierz (2014), the importance of lighting changes throughout the development of black cherry; at the seedling and sapling stages, growth of this species can be strongly reduced under the shade of a dense canopy. Importantly, in both natural and controlled conditions under shading net transmitting 10% of full light, black cherry seedlings may maintain green, photosynthetically active leaves even in the winter. This could also give this invader a competitive advantage over native tree species that shed their leaves in autumn (Bielinis et al. 2012). The costs and benefits of *P. serotina* maintaining some green leaves in the winter remain unclear, however.

Figure 1. A sessile oak seedling surrounded by mass natural regeneration of black cherry under a canopy of Scots pine (May 2012, Poznan University of Life Sciences, Zielonka Forest, Poland; photo credit: P. Robakowski).

Figure 2. Abundant natural regeneration of black cherry (1,200,000 seedlings per ha) and sessile oak seedlings growing in competition under a Scots pine canopy (May 2012, Poznan University of Life Sciences, Zielonka Forest, Poland; photo credit: P. Robakowski).

(a) (b)

Figure 3. One-year-old black cherry seedlings growing for six months (a) in full light and (b) under a shading net transmitting 10% of full light. The tallest seedling acclimated to full light was 5-fold taller than those grown in 10% light.

3.3. Temperature and Precipitation

On the Allegheny Plateau of Pennsylvania and New York, the center of the natural occurrence range of *P. serotina*, the climate is cool, moist and temperate, with an average annual precipitation of 970 – 1120 mm well distributed throughout the year. Summer precipitation averages 510 – 610 mm, and the frost-free growing season is 120 – 155 days in length. Winter snowfalls average 89 – 203 cm, and 45 – 90 days have a snow cover of 2.5 cm or more. Mean annual potential evapotranspiration is approximately 430 – 710 mm and mean annual water surplus is 100 – 610 mm. January temperatures average a maximum of 1 – 6°C and a minimum of -11 to -6°C, and July temperatures average a maximum of 27 – 29°C and a minimum of 11 – 16°C (Marquis 1990). In the introduced Central European range, the average annual precipitation ranges from 550 to 750 mm and the vegetative season lasts 180 – 210 days (mean 24-h temperature exceeds 5°C). Black cherry has adapted to a lower sum of precipitation than that found in its native North American range (Starfinger et al. 2003), which could be of great importance to the success of its invasion in Europe.

3.4. Soil Conditions

In Central Europe, *P. serotina* can grow within a wide range of forest sites, although it most commonly occurs in mixed coniferous and mixed deciduous forest types developed on rusty soils (podzols; Bijak et al. 2014). Black cherry finds optimal conditions for development on rusty soils and brown leached soils (Danielewicz 1994). This species shows low demands in terms of soil fertility, but it grows best on deep and fertile soils (Reinhart et al. 2003; Starfinger et al. 2003; Halarewicz 2011). At a young age, it is tolerant to strong acidity and low nutrient availability in soil (Halarewicz and Kawałko 2014). Black cherry density is correlated negatively with soil pH and positively with phosphorus content in upper soil layers (Starfinger et al. 2003). The leaves of *P. serotina* are rich in nitrogen, potassium, calcium and magnesium, although concentrations of these elements are significantly related to the light environment – higher leaf nutrient concentrations were observed in high light-acclimated seedlings than in shade-acclimated ones (Robakowski et al. 2016). In addition, degradation of *P. serotina* leaves is rapid and may occur over only three to four weeks (pot experiment, personal observations). These results confirm that the planting of black cherry characterized by low soil demands on poor sites provides nutrients through litter degradation and can be regarded as a semi-natural way of fertilizing the soil to increase the productivity of coniferous forests growing on poor sandy soils. According to Closset-Kopp et al. (2011), soil type significantly influences black cherry growth parameters (basal area and height) only when combined with light availability. Nevertheless, invasion rate and size are associated with the fact that black cherry populates faster and broader in areas with poor soils (podzols) than in areas with fertile soils (gleyed soils, luvisols or regosols). The species does not occur on moist soils in alder swamp forest and riparian forests, which could be explained by the fact that moist soils provide better development conditions for fungi of the genus *Pythium*. These pathogenic fungi are the natural enemies of *P. serotina* in its native range, but their importance in reducing black cherry abundance is lower in the introduced range (Halarewicz and Kawalko 2014).

4. Growth Strategy and Biomass Allocation

The growth of black cherry is similar to *R*-strategy, which also corresponds to its way of reproduction, lifetime, size of vegetative organs, etc. *R*-strategy consists of rapid growth, a short lifetime, production of many small seeds and the ability to pioneer (Grime 1977). Interestingly, however, *P. serotina* is able to switch from *K*- to *R*-strategy depending on its age. At the juvenile stage of development under a closed canopy, it develops *K*-strategy, with a very low growth rate, a long life span as an ‘Oskar’ and a strong re-sprouting capacity, following the ‘Alice behavior’ (Closset-Kopp et al. 2007) . At the maturity stage, however, it exhibits

R-strategy, characterized by a rapid growth rate, a short lifespan and abundant seed production, as described above. The ability to use *R* or *K* strategy allows *P. serotina* to overtop native tree species, giving it a competitive advantage over native tree species and contributing to its successful invasion (Closset-Kopp et al. 2011).

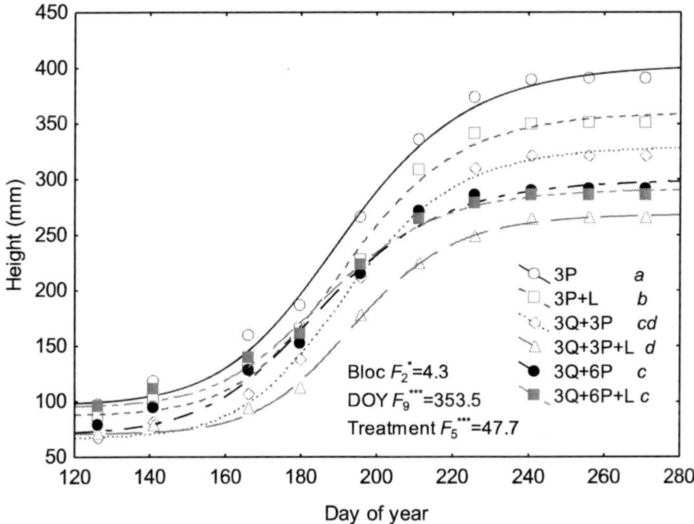

Figure 4. Height curves of *Prunus serotina* seedlings. Curves were obtained by fitting the mean values of seedling height with the sigmoid function $y = y_0 + a/1 + (T/x_0)^b$, where y = mean height of seedlings (n = 36), y_0 = the first (lowest) height value measured at DOY 126, T = day of year, and a, b and x_0 are the estimated parameters. The seedlings grew from April to October 2009 in one of six treatments: : 1) 3Q – three seedlings of *Quercus petraea*, 2) 3Q+L - three seedlings of *Q. petraea* + *Prunus serotina* leaves, 3) 3Q+3P - three seedlings of *Q. petraea* + three seedlings of *P. serotina*, 4) 3Q+3P+L - three seedlings of *Q. petraea* + three seedlings of *P. serotina* + *P. serotina* leaves, 5) 3Q+6P - three seedlings of *Q. petraea* + six seedlings of *P. serotina*, 6) 3Q+6P+L - three seedlings of *Q. petraea* + six seedlings of *P. serotina* + *P. serotina* leaves. The results of the main effects ANOVA with block, day of year (DOY) and treatment as the sources of variance are given in a plot. F = the value of Snedecor's function with the degrees of freedom in lower case; $^{***}P < 0.001$, $^{**}0.001 \leq P < 0.01$, $^{*}0.01 \leq P < 0.05$. The same letters indicate that the treatments were not statistically significantly different based on Tukey's test at $\alpha = 0.05$ (Robakowski and Bielinis 2011).

Black cherry seedlings grow rapidly in the nursery, gaining 46 or even 91 cm in height per year. If fertilized, they can grow 1.2 – 1.8 m in a year or, in some exceptional cases, even up to 2.4 m (Marquis 1990). In its introduced range, *P. serotina* can exhibit faster growth dynamics than those of native tree species. *Prunus serotina* has been called 'the new ecosystem engineer,' as it promotes traits that enable species to capture resources in the new environment it creates by invading, reducing heterogeneity of the environment and diversity of species (Closset-Kopp et al. 2011). In a controlled experiment, 1-year-old seedlings showed a continuous and steep height increase over the whole vegetative season. When the mean initial heights in April were compared with those after the bud set in September, cherry seedlings were 4-fold taller in September (Figure 4; Robakowski and Bielinis 2011). The growth dynamics of cherry were rapid from the beginning up to the end of the vegetative season, outstripping seedlings of sessile oak, its native slow-growing competitor. An investigation of cherry height growth and the ratio of height change to final height over the whole vegetative period confirmed the opinion that its rapid growth is a great advantage for this invasive species when growing in competition with native species (Marquis 1990; Starfinger 2003). In a mid-shade pot experiment conducted in October, at the end of the vegetative season, black cherry was on average 1.5-fold taller than sessile oak (Robakowski and Bielinis 2011). Its rapid height growth gave cherry seedlings an advantage over oak seedlings when competing for light. As a moderately shade-tolerant tree, black cherry does not prefer shading from neighbors in the brighter light environment and tends to overtop them (Halarewicz and Żołnierz 2014). Another pot experiment indicated that cherry can beat oak in the competition for light under different pressures of intra- and interspecific competition, although mulching with fresh cherry leaves and its responses may have been modulated by the light environment during growth (Robakowski et al. 2018).

Dynamic growth rates are particularly important when *P. serotina* competes with native forest trees in its introduced range. Relative growth rates (RGR) and biomass allocation of invasive *P. serotina* and native *Quercus petraea* were compared in a pot experiment using different combinations of seedlings with or without mulching of fresh *P. serotina* leaves (Robakowski et al. 2018). Two-year-old *P. serotina* seedlings had higher RGR than did *Q. petraea* (0.133 ± 0.004 and 0.09 ± 0.004 g total dry mass day^{-1}, respectively; mean ± SE).

(a)

(b)

Figure 5. Biomass allocation in seedlings of *Prunus serotina* (a.) and *Quercus petraea* (b) acclimated to 100%, 25% or 10% of full sunlight from May to October 2012 (Photo credit: P. Robakowski and E. Bielinis).

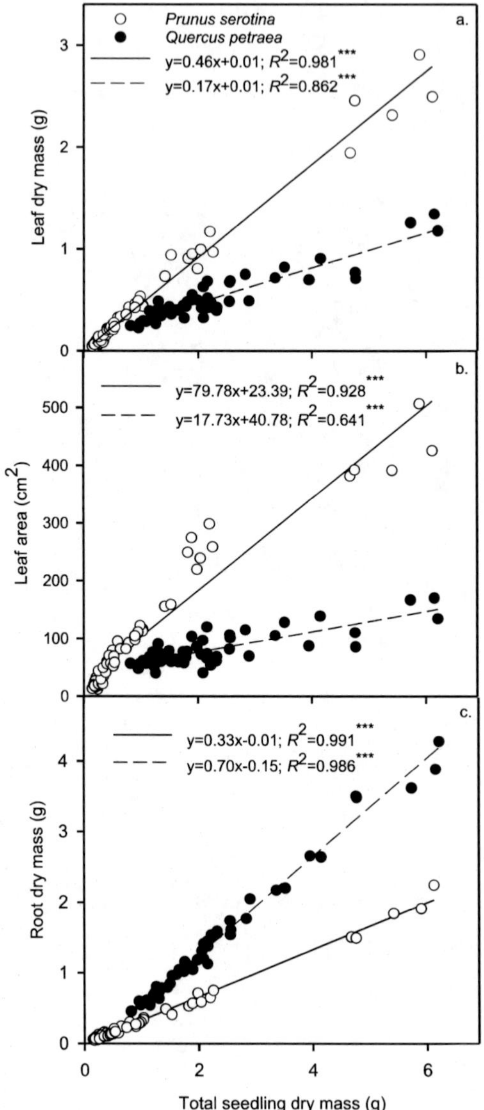

Figure 6. Linear regressions of (a) leaf dry mass, (b) leaf area or (c) root dry mass on total seedling dry mass. Seedlings were grown in one of three light treatments (10%, 25% or 100% of full sunlight) and three combination treatments (P – three *Prunus serotina* seedlings, Q – three *Quercus petraea* seedlings, or P+Q+L – six *P. serotina* + three *Q. petraea* + mulching with *P. serotina* fresh leaves). Each point represents the mean value of three seedlings per species per pot ($n = 216$ seedlings per species; Robakowski et al. 2018).

Furthermore, the differences between the species increased as light availability increased: compared to *Q. petraea*, RGR of *P. serotina* was 1.8-fold higher in low light (LL; 10% of full sunlight), 2.1-fold higher in medium light (ML; 25%) and 2.3-fold higher in high light (HL; 100%). RGR of *Q. petraea* did not differ significantly among competition and mulching treatments, whereas RGR of *P. serotina* seedlings growing in the presence of interspecific competition and mulching in low light were greater than those of *P. serotina* growing in monoculture or in competition with *Q. petraea*. Leaf area ratio (LAR) decreased and total leaf area (AL) increased with increasing light for both species, but more significantly for the invasive than native species. In LL, RGR, LAR and AL were highest in *P. serotina* growing with *Q. petraea*, whereas in HL, RGR and AL were lower in this combination than in the interspecific competition with mulching (Figure 5). Interestingly, at the same total biomass, root dry mass was lower, and AL and total leaf dry mass were higher, in seedlings of *P. serotina* than in seedlings of *Q. petraea* (Figure 6; Robakowski et al. 2018). The investment in above-ground organs, especially in total AL and mass, contributes to an enhancement of light capture and use in photosynthesis. Biomass allocation to organs is crucial in interspecific competition and can play an important role in the invader's strategy. Conversely, the opposite biomass allocation strategy of sessile oak allows that species to sometimes successfully compete with black cherry in natural conditions under the Scots pine canopy.

5. Leaf Structure and Gas Exchange

The global leaf economic spectrum indicates that species with lower leaf mass per area (LMA) have greater photosynthetic capacities (Wright et al. 2004). In a pot experiment, the mean LMA value of *P. serotina* (43 ± 1 g m^{-2}) outside of its natural geographic range was lower than the LMA of this species in Minnesota, USA, where it naturally occurs (48 ± 3 g m^{-2}; Sendall and Reich 2013; Robakowski et al. 2018). A comparison of functional traits of *P. serotina* leaves between its native range across

Eastern North America (New York) and its invasive range in northern France showed similar maximal net CO_2 assimilation rates (A_{max}), but invasive populations invested more carbon to leaves, showing higher dark respiration, higher leaf carbon concentration and lower LMA than in the native range (Heberling et al. 2016). This suggests that invasive *P. serotina* is able to decrease LMA in its non-native geographical range. *Prunus serotina* exhibits increased LMA in full sunlight (100%) when compared with low (10%) and mid light (25%; Robakowski et al. 2018). Pattison et al. (1998) showed that five invasive and four native species grown in the shade had lower LMA than those grown in full sun. In contrast, other studies have shown that *P. serotina* juveniles growing in full sun have approximately 3-fold higher LMA than in the understory (Harrington et al. 1989; Sendall and Reich 2013). Leaf structural changes reflected by LMA, together with biochemical mechanisms of excess energy dissipation, allow plants to use light more efficiently for photosynthesis and protect the photosynthetic apparatus against excess energy (Ellsworth and Reich 1992; Niinemets et al. 1998).

Prunus serotina and *Q. petraea* both increase A_{max}, photosynthetic nitrogen use efficiency (PNUE) and LMA with increasing light (Robakowski et al. 2018). High light availability, however, favors *P. serotina* over *Q. petraea*, supporting the hypothesis that the ability of *P. serotina* to suppress native *Q. petraea*, and likely other native tree species, can be enhanced under greater resource supply. Both competition with *P. serotina* and mulching with *P. serotina* leaves decreased the LMA and A_{max} of *Q. petraea* when compared with *Q. petraea* monoculture with mulching, which indicates the negative effect of interspecific competition on *Q. petraea* leaf structure and photosynthesis. Interestingly, in 25% of full light, an increase in prunasin concentration in black cherry roots is associated with higher A_{max} and PNUE, which suggests a trade-off between increasing assimilation due to higher PNUE and greater prunasin concentration (Robakowski et al. 2016).

6. Why Has the *Prunus serotina* Invasion Been Successful?

Black cherry's ability to achieve invasive success over such a large area of European forests is fairly unique among non-native woody species. Alien plant invasions result from a complex interaction between the species' life traits and attributes of the recipient ecosystem (Closset-Kopp et al. 2007). In the introduced range in European countries, *P. serotina* has been considered an aggressive invader and gained the reputation of being a 'pest' (Starfinger et al. 2003). The success of invasion can to some extent be explained by the enemy release hypothesis, which states that invasive species escape from natural enemies and develop novel competitive abilities in a new environment (Keane and Crawley 2002). There is evidence that the invasiveness of *P. serotina* in Europe is due in part to its escape from *Pythium* spp., a primary competitor in its native range (Reinhart et al. 2003). Observations in European forests and the results of a pot experiment with different combinations of competing seedlings of *P. serotina* and *Q. petraea* suggest, however, that the resource-enemy release hypothesis (Blumenthal 2005), in which the enemy release hypothesis and increased resource availability act in concert, may more accurately explain the invasion of *P. serotina* in Europe (Robakowski et al. 2016, 2018).

In its introduced range, *P. serotina* competes with native tree species. To simulate the competition naturally occurring between seedlings of *Q. petraea* and *P. serotina* in the forest, a controlled experiment was conducted using 1-year-old seedlings of both species (Robakowski and Bielinis 2011). Eight treatments at different competition intensities regulated the number of competing seedlings. The treatments were established by varying the number of potted seedlings and adding fresh cherry leaves to the substrate to enhance allelopathic effects. After one vegetative season, no inhibitory effect on oak growth was observed. Instead, the presence of cherry seedlings enhanced oak height growth. This could be due to either the strong interspecific competition for light or, less plausibly, a positive allelopathic effect, or an interaction of both. In this experiment, there was also a negative, or autotoxic, effect of cherry seedlings and/or mulching with fresh cherry leaves on the height of cherry seedlings.

Generally, observations have indicated that *P. serotina* does not have many enemies in central European forests (Starfinger et al. 2003). This could be at least partly due to a presence of the cyanogenic glycoside prunasin in the leaves, shoots and roots of black cherry, as well as amygdalin in its seeds (Vetter 2000). These allelochemicals make plant tissues less attractive for feeding by insects and ungulates, while also inhibiting respiration, photosynthesis and growth of other competing plants. In Poland, in forests with numerous populations of red deer and roe deer, black cherry seedlings are not browsed except for slight damage of young shoots in early spring (personal observations). After being hydrolyzed by endogenous enzymes, prunasin decomposes into hydrogen cyanide (HCN), a respiratory poison that inhibits the activity of metalloenzymes, including cytochrome c oxidase (Leavesley et al. 2008). Highly toxic to aerobic organisms, HCN inhibits respiratory function in cells, and thus impedes seed germination and plant growth rates (Gleadow and Møller 2014). Leachates of *P. serotina* leaves containing prunasin and other potentially allelochemical compounds have been shown to inhibit seed germination and growth of *Salix alba* (L.) seedlings (Csiszár et al. 2013). HCN is produced when plant tissue has been damaged, which suggests that prunasin may defend cyanogenic plants against herbivores (Swain and Poulton 1994; Ballhorn et al. 2011; Agrawal et al. 2012). For grazing animals, daily doses in excess of 50 mg HCN/kg body weight are considered dangerous (Majak et al. 1980).

Allelopathy is defined as the chemical influence of volatile, leached compounds or exudates from different parts of plants that can reduce or enhance growth, reproduction and/or physiological processes of neighboring organisms – not only plants, but also mycorrhizal fungi, bacteria, etc. (Rizvi et al. 1992; Weir 2007). Allelochemical activity can give an invasive species an advantage over native species (Callaway and Aschehoug 2000; Hierro and Callaway 2003). In an introduced environment, an invader may use a ‘new weapon’ to compete more efficiently with native species (Novoplansky 2009; Yuan et al. 2013). Prunasin, as an allelochemical in black cherry tissues, may inhibit growth of competing plants, as well as being a repellent against herbivores.

The concentration of prunasin in cherry roots increases from May up to its maximum value in August, concurrent with air temperature and *P. serotina* growth rate, and then declines in September (Robakowski et al. 2018). This observation suggests that in a warmer climate, prunasin may not only play the role of an effective deterrent against insects or browsers when accumulated in roots, twigs and leaves (Patton et al. 1997; Ballhorn 2011), but may also act as a carbon store (Selmar et al. 1988). Changes in the concentration of cyanogenic glycosides affect the balance of carbon and nitrogen in a plant; hence, prunasin and amygdalin may take part in carbon and nitrogen cycles as their sink or source. Higher prunasin concentration in low light may be due to the fact that shade-acclimated plants are more severely threatened by damage caused by insects, possibly causing them to invest more resources into secondary metabolites (Karolewski et al. 2010, 2013). Interspecific competition between cherry and sessile oak, and mulching with fresh cherry leaves, does not appear to stimulate prunasin concentration in *P. serotina* roots. In full sun, however, the concentration of prunasin in roots of potted, monoculture *P. serotina* scaled with growth, biomass allocation and photosynthesis. Under more favorable light conditions, the resources stored in roots as prunasin may be transformed and transported to leaves and reused for photosynthesis and biomass production. This is in agreement with observations of *P. serotina* seedlings that initially grow slowly or stagnate under a dense canopy of young Scots pine forest, whereas under a more open canopy their growth dynamics are enhanced several-fold. This suggestion is also in agreement with resource allocation-based hypotheses regarding prunasin costs and how multifunctional attributes of prunasin may lower the cost of interspecific competition (Neilson et al. 2013).

In forests, allelopathy often overlaps with other interactions between plants and its effects may be difficult to distinguish from those of the competition for light, water and nutrients, or other plant–plant and plant–animal interactions. The results of a pot experiment with black cherry and sessile oak seedlings did not support the hypothesis that cherry had an inhibitory effect on oak growth, at least after one vegetative season. Instead, the presence of cherry seedlings enhanced oak height growth rates, which could be due to either the strong interspecific competition for light or, less plausibly, a positive allelopathic effect, or an interaction of both. Interestingly, the presence of other cherry seedlings and/or fresh cherry leaves used for mulching had an autotoxic effect on the height of cherry seedlings (Robakowski and Bielinis 2011).

The competition of native forest trees for light and allelochemical interactions with an invasive species at an early age has an important ecological, evolutionary and practical role in silviculture (Le Duc 1998; Balandier et al. 2006). The natural selection of tree species can be strongly impacted by a fast-growing and easily reproducing invader gaining an advantage in competition for site resources and equipped with a new allelochemical weapon, as in case of *P. serotina*.

There is no evidence that seedlings of black cherry inhibit the growth of sessile oak. In contrast, in a pot experiment under optimal irrigation and fertilization, cherry seedlings competing for light appeared to force oak seedlings to gain height at a higher rate. Black cherry may play a role of filler towards oak, increasing its height growth when growing in competition, although a positive allelopathic effect cannot be entirely excluded (Rizvi et al. 1992). In contrast to oak seedlings, the height and diameter at root collar of black cherry seedlings decreases with greater competition. It remains unclear, however, if there is an effect of interspecific or intraspecific interaction, or both. A lower height reached by cherry seedlings grown in monospecific combination (three cherry seedlings per pot) in comparison to those mulched with *P. serotina* leaves suggests that an autotoxic effect may diminish the growth of cherry seedlings (Robakowski and Bielinis 2011).

Interestingly, when *P. serotina* grows in competition with *Q. petraea*, the effect of light environment on leaf-absorbed light energy partitioning in three competing processes – (1) photosynthesis (Φ_{PSII}), quantum yield of PSII photochemistry; (2) energy dissipation as heat (Φ_{NPQ}), quantum yield of ΔpH- and xanthophyll-regulated thermal energy dissipation and (3) constitutive energy dissipation as heat and fluorescence ($\Phi_{f,D}$) – was more pronounced than the effects of interspecific competition and allelopathy (Robakowski et al. 2018). It is worth noting that *P. serotina* increases Φ_{NPQ} and Φ_{PSII} markedly at the expense of $\Phi_{f,D}$ when growing in 25% and 100% full sunlight, in comparison with low light of 10%, especially when in competition with *Q. petraea* with mulching. This response of invasive *P. serotina* to light can be regarded as an 'Oskar syndrome' at the bioenergetics level (Closset-Kopp et al. 2007, 2011). Larger amounts of leaf-absorbed light energy translocated to Φ_{PSII} can contribute to an increase in growth under optimal light conditions, and greater plasticity of light energy partitioning can give an important competitive advantage to invasive over native species.

In forests infested by black cherry, there are more plant species that require high levels of nitrogen supply, soil reaction and moisture content. A negative correlation has been found between the abundance of *P. serotina* and the diversity of the understory species composition, although the effect of this invasive species on plant species diversity of the forest floor is overridden by variable micro-habitat conditions (Halarewicz and Żołnierz 2014). Community changes do occur following invasion, but understory species richness is only reduced in heavily invaded stands. Under natural conditions, it is difficult to disentangle the invader's biological effects from a simple density effect on native, non-invasive species (Verheyen et al. 2007).

7. *Prunus serotina* and Global Climate Change

The successful invasion of black cherry in European countries suggests that this species is able to adapt easily to a new environment. In its invasive range, *P. serotina* has developed a different resource-use strategy than within its native range, but whether this divergence is due to species-specific plasticity or adaptation to the site conditions has not yet been determined (Heberling et al. 2016). The fact that in Central Europe *P. serotina* grows under lower annual precipitation than in its natural occurrence range indicates that it possesses greater plasticity to adapt to global warming than native species characterized by a lower invasiveness. Climatic models of changes in *P. serotina* distribution predict the disappearance of suitable sites, however, which may cause possible inbreeding and gene drift that will result in the loss of genetic diversity of the species. Modelling not only indicates that currently suitable areas may disappear due to global warming, but also predicts that additional areas may arise where climatic conditions will be favorable for black cherry. The larger proportion of these new areas is concentrated in the northern latitudes of the United States. In Mexico, new areas will be at higher altitudes, implying altitudinal displacement of the species (Seguera et al. 2018).

Morphological and physiological variability can reflect the adaptation of a species to different habitats, thus increasing their probability of surviving in the context of climate change. *Prunus serotina* has been shown to adapt to different climatic conditions; therefore, it is possible that its genetic variability may allow it to adapt to future climate conditions. The genetic diversity of *P. serotina*, represented by different subspecies and ecotypes, allows this species to adapt to environmental changes (Hughes and Stachowicz 2004). Its invasive ability in distinct geographic regions worldwide (Starfinger et al. 2003) and its capacity to adapt to a broad range of climates support a positive outlook for *P. serotina* in the future climate. Although the most current models of species distribution indicate that climate is the main factor that will determine changes in the geographical ranges of plant species, other factors such as soil conditions, competition, allelopathy and plant–animal interactions also affect the responses of any species to global climate change. When genetic structures of invasive and native *P. serotina* populations were compared, the overall genetic structure displayed by microsatellite loci was more pronounced in the invasive range than in the native range, with a reduction in the overall number of alleles, allelic richness and heterozygosity found at microsatellite loci in the invasive range (Pairon et al. 2010). Genetic diversity and genetic markers related to the ability to adapt to a changing climate should be combined with phenotypic patterns in the existing climatic models in both the native and invasive ranges to predict changes in the geographical distribution of *P. serotina* at a higher confidence level than that offered by existing models.

ACKNOWLEDGMENTS

This publication is co-financed within the framework of the Ministry of Science and Higher Education program as "Regional Initiative Excellence" in the years 2019 – 2022, project number 005/RID/2018/19.

REFERENCES

Agrawal, A. A., Hastings, A. P., Johnson, M. T. J., Maron, J. L., Salminen, J. (2012) Insect herbivores drive real-time ecological and evolutionary change in plant populations. *Science* 338: 113–116.

Auclair, A. N. and Cottam, G. (1971) Dynamics of black cherry (*Prunus serotina* Ehrh.) in southern Wisconsin oak forests. *Ecological Monographs* 41: 153–177.

Balandier, P., G. Sonohat, H. Sinoquet, C. Varlet-Grancher, and Dumas, Y. (2006) Characterisation, prediction and relationships between different wavebands of solar radiation transmitted in the understorey of even-aged oak (*Quercus petraea*, *Q. robur*) stands." *Trees* 20 (3): 363–70. https://doi.org/10.1007/s00468-006-0049-3.

Ballhorn, D. J. (2011) Constraints of simultaneous resistance to a fungal pathogen and an insect herbivore in lima bean (*Phaseolus lunatus* L.). *Journal of Chemical Ecology* 37 (2): 141–44. https://doi.org/10.1007/s10886-010-9905-0.

Ballhorn, D. J., Kautz, S., Jensen, M., Schmitt, I., Heil, M. and Hegeman, A. D. (2011) Genetic and environmental interactions determine plant defences against herbivores. *Journal of Ecology* 99 (1): 313–26. https://doi.org/10.1111/j.1365-2745.2010.01747.x.

Bielinis, E., Robakowski, P., Wyka, T., Samardakiewicz, S. (2012) Why does black cherry maintain some leaves during the winter? *Acta Physiologiae Plantarum*, 38 (Suppl 1): 86.

Bijak, Sz., Czajkowski, M. and Ludwisiak, Ł. (2014) Occurrence of black cherry (*Prunus serotina* Ehrh.) in the State Forests in Poland. *Forest Research Papers* 75 (4): 359–65. https://doi.org/10.2478/frp-2014-0033.

Blumenthal, D. (2005) Interrelated causes of plant invasion. *Science* 310: 243–44.

Callaway, R. M. and Aschehoug, E. T. (2000) Invasive plants versus their new and old neighbors: a mechanism for exotic invasion. *Science* 290: 521–23. https://doi.org/10.1126/science.290.5491.521.

Csiszár, Á. (2009) Allelopathic effect of invasive woody plant species in Hungary. *Acta Silvatica et Lignaria Hungarica* 5: 9–17.

Closset-Kopp, D., Chabrerie, O., Valentin, B., Delachapelle, H., and Decocq, G. (2007) When Oskar meets Alice: does a lack of trade-off in r/K-strategies make *Prunus serotina* a successful invader of European forests?" *Forest Ecology and Management* 247 (1–3): 120–30. https://doi.org/10.1016/j.foreco.2007.04.023.

Closset-Kopp, D., Saguez, R. and Decocq, G. (2011) Differential growth patterns and fitness may explain contrasted performances of the invasive *Prunus serotina* in its exotic range. *Biological Invasions* 13 (6): 1341–55. https://doi.org/10.1007/s10530-010-9893-6.

Csiszár, Á., Korda, M., Schmidt, D., Šporčič, D., P. Süle, and B. Teleki et al. 2013. Allelopathic potential of some invasive plant species occurring in Hungary. *Allelopathy Journal* 31(2): 309–18.

Danielewicz, W. (1994) Rozsiedlenie czeremchy amerykanskiej (*Prunus serotina* Ehrh.) na terenie Nadlesnictwa Doswiadczalnego Zielonka. ["Distribution of black cherry (*Prunus serotina* Ehrh.) in the Experimental Forest 'Zielonka'] *Prace Komisji Nauk Rolniczych i Komisji Nauk Lesnych, Poznanskiego Towarzystwa Przyjaciól Nauk, Wydzialu Nauk Rolniczych i Lesnych* 78: 35–42. (in Polish).

Deckers, B., Verheyen, K., Hermy, M. and Muys, B. (2005) Effects of landscape structure on the invasive spread of black cherry *Prunus serotina* in an agricultural landscape in Flanders, Belgium. *Ecography* 28 (1): 99–109. https://doi.org/10.1111/j.0906-7590.2005.04054.x.

Ellsworth, D. S., Reich, P. B. (1992) Leaf mass per area, nitrogen content and photosynthetic carbon gain in *Acer saccharum* seedlings in contrasting forest light environmnents. *Functional Ecology* 6: 423–435.

Gleadow, R. M., and Møller, B. L. (2014) Cyanogenic glycosides: synthesis, physiology, and phenotypic plasticity. *Annual Review of Plant Biology* 65: 155–85. https://doi.org/10.1146/annurev-arplant-050213-040027.

Grime, P. J. (1977) Evidence for the existence of three primary strategies in plants and its relevance to ecological and evolutionary theory. *American Naturalist* 111: 1169–1194.

Guzmán, F. A., Segura, S., Fresnedo-Ramírez, J. (2018) Morphological variation in black cherry (*Prunus serotina* Ehrh.) associated with environmental conditions in Mexico and the United States. *Genetic Resources and Crop Evolution* 65(8): 2151–2168.

Halarewicz, A. (2011) The Reasons underlying the invasion of forest communities by black cherry, *Prunus serotina* and its subsequent consequences. *Forest Research Papers* 72 (3): 267–72. https://doi.org/10.2478/v10111-011-0026-5.

Halarewicz, A., Kawałko, D. (2014) Wpływ czynników glebowych na występowanie *Prunus serotina* w fitocenozach leśnych. [Effect of soil factors on the incidence of *Prunus serotina* in forest phytocoenoses]. *Sylwan* 158(2): 117−123.

Halarewicz, A., Żołnierz, L. (2014) Changes in the understorey of mixed coniferous forest plant communities dominated by the American black cherry (*Prunus serotina* Ehrh.). *Forest Ecology and Management* 313: 91–97.

Harrington, R. A., Brown, B. J., Reich, P. B. and Fownes, J. H. (1989) Ecophysiology of exotic and native shrubs in southern Wisconsin. *Oecologia* 80(3): 368–73. https://doi.org/10.1007/bf00379038.

Heberling, M. J., Kichey, T., Decocq, G., and Fridley, J. D. (2016) Plant fuctional shifts in the invaded ranges: a test with reciprocal forest invaders of Europe and North America. *Functional Ecology* 30: 875–884. https://doi.org/10.1111/1365-2435.12590

Hierro, J. L., and Callaway, R. M. (2003) Allelopathy and exotic plant invasion. *Plant and Soil* 256 (1): 29–39. https://doi.org/10.1023/A:1026208327014.

Hough, R. B. (1957) *Encyclopedia of American Woods*. R. Speller and Sons, New York.

Hughes, A. R., Stachowicz, J. J. (2004) Genetic diversity enhances the resistance of a seagrass ecosystem to disturbance. *Proceedings of the National Academy of Sciences* 101(24): 8998–9002.

Karolewski, P., Zadworny, M., Mucha, J., Napierała-Filipiak, A., and Oleksyn, J. 2010. Link between defoliation and light treatments on root vitality of five understory shrubs with different resistance to insect herbivory. *Tree Physiology* 30:969–78.

Karolewski, P., Giertych, M. J., Żmuda, M., Jagodziński, A. M. and Oleksyn, J. (2013) Season and light affect constitutive defenses of understory shrub species against folivorous insects. *Acta Oecologica* 53: 19–32. https://doi.org/10.1016/j.actao.2013.08.004.

Keane, R. M., and Crawley, M. J. (2002) Exotic plant invasions and the enemy release Hypothesis. *Trends in Ecology and Evolution* 17(4): 164–70. https://doi.org/10.1016/S0169-5347(02)02499-0.

Leavesley, H. B, Li L., Prabhakaran, K., Borowitz, J. L. and Isom, G. E. (2008) Interaction of cyanide and nitric oxide with cytochrome c oxidase: implications for acute cyanide toxicity. *Toxicological Sciences : An Official Journal of the Society of Toxicology* 101 (1): 101–11. https://doi.org/10.1093/toxsci/kfm254.

Le Duc, M. G. and D. C. Hawill (1998) Competition between *Quercus petraea* and *Carpinus betulus* in an ancient wood in England: seedling survivorship. *Journal of Vegetation Science* 9: 873–880.

Majak, W., Quinton, D. A., Broersma, K. (1980) Cyanogenic glycoside levels in saskatoon serviceberry. *Journal of Range Management* 33 (3): 197–99.

Marquis, D. A. (1990). *Prunus serotina* Ehrh. Black Charry. In: Burns RM, Honkala BH, technical co-ordinators. *Silvics of North America.*

Volume 2. *Hardwoods. Agriculture Handbook 654*. Washington, DC: United States Department of Agriculture, Forest Service. https://www.srs.fs.usda.gov/pubs/misc/ag_654/volume_2/silvics_v2.pdf.

Möllerová, J. (2005) Notes on invasive and expansive trees and shrubs. *Journal of Forest Science* 51: 19–23.

Neilson, E. H., Jason, Q., Goodger, D. Woodrow, I. E. and Møller, B. L. (2013) Plant chemical defense: at what cost? *Trends in Plant Science* 18 (5): 250–58. https://doi.org/10.1016/j.tplants.2013.01.001.

Niinemets, Ü., Bilger, W., Kull, O., Tenhunen, J. D. (1998) Acclimation to high irradiance in temperate deciduous trees in the field: changes in xanthophyll cycle pool size and in photosynthetic capacity along a canopy light gradient. *Plant, Cell & Environment* 21: 1205–1218.

Novoplansky, A. (2009) Picking battles wisely: plant behaviour under competition. *Plant, Cell & Environment* 32 (6): 726–41. https://doi.org/10.1111/j.1365-3040.2009.01979.x.

Pairon, M., Petitpierre, B., Campbell, M., Guisan, A., Broennimann, O., Baret, P. V., A. L. Jacquemart and Besnard, G. (2010) Multiple introductions boosted genetic diversity in the invasive range of black cherry (*Prunus serotina*; *Rosaceae*). *Annals of Botany* 105: 881–890.

Pattison, R. R., Goldstein, G., Ares, A. (1998) Growth, biomass allocation and photosynthesis of invasive and native Hawaiian rainforest species. *Oecologica* 117: 449–59.

Patton, C. A., Ranney, T. G., Burton, J. D., Walgenbach, J. F. (1997) Natural pest resistance of *Prunus* taxa to feeding by adult Japanese beetles: role of endogenous allelochemicals in host plant resistance. *Journal of American Society of Horticultural Sciences* 122 (5): 668–72.

Reinhart, K. O., Packer, A., Van der Putten, W. H. and Clay, K. (2003) Plant-soil biota interactions and spatial distribution of black cherry in its native and invasive ranges. *Ecology Letters* 6 (12): 1046–50. https://doi.org/10.1046/j.1461-0248.2003.00539.x.

Rizivi, S. J. H., Haque, H., Singh, V. K., Rizivi, V. (1992) A discipline called allelopathy. In *Allelopathy. Basic and applied aspects*, edited by: S. J. H. Rizivi, V. Rizivi. Chapman & Hall, pp. 1–8.

Robakowski, P. and Bielinis, E. (2011) Competition between sessile oak (*Quercus petraea*) and black cherry (*Padus serotina*): dynamics of seedlings growth. *Polish Journal of Ecology* 59 (2).

Robakowski, P., E. Bielinis, and Sendall, K. (2018) Light energy partitioning, photosynthetic efficiency and biomass allocation in invasive *Prunus serotina* and native *Quercus petraea* in relation to light environment, competition and allelopathy. *Journal of Plant Research* 131(3): 505–523. https://doi.org/10.1007/s10265-018-1009-x.

Robakowski, P., Bielinis, E., Mejza, I., Bułaj, B. (2016) Seasonal changes affect root prunasin concentration in *Prunus serotina* and override species interactions between *P. serotina* and *Quercus petraea*. *Journal of Chemical Ecology* 42(3): 202-214. doi 10.1007/s10886-016-0678-y.

Robakowski, P., Stachnowicz, W. (2000) Centre for Agriculture and Biosciences International 2000. *Padus serotina*. Robakowski, P., Stachnowicz, W. In: Forestry Compendium, Global Edition. Wallingford, UK: CAB International.

Segura, S., Guzmán-díaz, F., López-upton, J. and Mathuriau, C. (2018) Distribution of *Prunus serotina* Ehrh. in North America and its invasion in Europe. 6: 111–124. https://doi.org/10.4236/gep.2018.69009.

Selmar, D., Lieberei, R., and Biehl, B. (1988) Mobilization and utilization of cyanogenic glycosides: the linustatin pathway. *Plant Physiology* 86: 711–16. https://doi.org/10.1104/pp.86.3.711.

Sendall, K., Reich, P. B. (2013) Variation in leaf and twig CO_2 flux as a function of plant size: a comparison of seedlings, saplings and trees. *Tree Physiology* 33: 713-729.

Silvertown, J. (1982) *Introduction to Plant Population Ecology*. Longman, London.

Smith, A. J. (1975) Invasion and ecesis of bird-disseminated woody plants in a temperate forest sere. *Ecology* 56(1): 19-34.

Starfinger, U. (1997) Introduction and naturalization of *Prunus serotina* in central Europe. In: Brock, J. H., Wade, M., Pysek, P. and Green, D.

(eds) *Plant Invasions: Studies from North America and Europe*, pp. 161–171. Backhuys, Leiden, The Netherlands.

Starfinger, U., Kowarik, I., Rode, M. and Schepker, H. (2003) From desirable ornamental plant to pest to accepted addition to the flora? - The perception of an alien tree species through the centuries. *Biological Invasions* 5 (4): 323–35. https://doi.org/10.1023/B:BINV.0000005573.14800.07.

Swain, E., Poulton, J.E. (1994) Utilization of amygdalin during seedling development of *Prunus serotina. Plant Physiology* 106: 437–445.

Verheyen, K., Vanhellemont, M., Stock, T. and Hermy, M. (2007) Predicting patterns of invasion by black cherry (*Prunus serotina* Ehrh.) in Flanders (Belgium) and its impact on the forest understorey community. *Diversity and Distributions* 13 (5): 487–97. https://doi.org/10.1111/j.1472-4642.2007.00334.x.

Vetter, J. (2000) Plant cyanogenic glycosides. *Toxicon* 38 (1): 11–36. https://doi.org/10.1016/S0041-0101(99)00128-2.

Weir, T. L. (2007) The role of allelopathy and mycorrhizal associations in biological invasions. *Allelopathy Journal* 20(1): 43–50. https://doi.org/citeulike-article-id:5726087.

Wendel, G. W. (1975) Stump sprout growth and quality of several Appalachian hardwood species after clearcutting. *USDA Forest Service, Research Note NE-239*. Northeastern Forest Experiment Station Upper Darby, PA. https://www.fs.fed.us/ne/newtown_square/publications/research_papers/pdfs/scanned/OCR/ne_rp329.pdf (2019-11-25).

Wright, I. J, Reich, P. B., Westoby, M., Ackerly, D. D., Baruch, Z., Bongers, F., Cavender-bares, j. et al. (2004) The worldwide leaf economics spectrum. *Nature* 428: 821–27.

Yuan, Y., Weihua, G., Wenjuan, Ding, and Wang, R. (2013) Competitive interaction between the exotic plant rhus typhina L. and the native tree *Quercus acutissima* Carr. in northern China under different soil N : P Ratios. *Plant Soil* 372: 389–400. https://doi.org/10.1007/s11104-013-1748-3.

BIOGRAPHICAL SKETCH

Piotr Robakowski

Affiliation: Poznan University of Life Sciences, Faculty of Forestry

Education:

2006 Habilitation: The August Cieszkowski Agricultural University of Poznań, Faculty of Forestry. Field of study: plant ecophysiology. The thesis for post-doctoral degree topic: "Ecophysiology of silver fir (*Abies alba* Mill.) at the young age."

1997 Doctorate: The August Cieszkowski Agricultural University of Poznań, Department of Forestry. Field of study: plant ecophysiology. Doctoral dissertation topic: "Impact of ultraviolet-B radiation on selected provenances of some species of forest trees and vegetal cover". The research work was conducted at the Gembloux Agricultural Faculty, Belgium in the frame of the "Individual Mobility Grants of European Union "Tempus" 1993/1994, 1996.

1991 Graduate studies: The August Cieszkowski Agricultural University of Poznań, Faculty of Forestry. Field of study: Forestry. The thesis for a master's degree topic: "Variability of the soil arthropods in the stands being under an impact of gaseous emissions from the Chemical Factory "Police".

Business Address: Poznan University of Life Sciences, Faculty of Forestry, Wojska Polskiego 71E, PL 60-625 Poznan, Poland

Research and Professional Experience, Professional Appointments:

2008 – Associate and full Professor, Poznan University of Life Sciences, Faculty of Forestry, Chair of Site Sciences and Forest Ecology, Ecological Foundation of Silviculture Unit. Lectures and classes in forest ecology, pro-ecological sylviculture, global climate changes. Lectures and classes in forest ecophysiology (in English).

2006 - 2007 Associate professor, the August Cieszkowski Agricultural University of Poznań, Faculty of Forestry, Department of Silviculture, Ecological Foundation of Silviculture Unit. Lectures and classes in ecological rudiments of silviculture, lectures on global change, monographic lecture on effects of global changes on forests in English, seminars and theses for a master's degree, research on plant ecophysiology.

1998 – 2006 Assistant professor, the August Cieszkowski Agricultural University of Poznań, Faculty of Forestry, Department of Silviculture, Ecological Basis of Silviculture Unit. Lectures and classes in ecological basis of silviculture, monographic lecture on global changes (the program "Socrates"), seminars and theses for a master's degree, research on forest ecology and ecophysiology.

1995 – 1998 Tutor in ecological basis of silviculture, a doctor of forest sciences, the August Cieszkowski Agricultural University of Poznań, Department of Forestry. Preparation of doctoral dissertation, classes, seminars and theses for master's degree, research on forest ecophysiology, evaluation of spruce defoliation in the Sudety Mountains (Southern Poland).

1991 – 1994 Technician at the Institute of Dendrology, Polish Academy of Sciences, Kórnik. Research work in the laboratory of tissue culture (one year): vegetative multiplication of Syringa sp. and Rhododendron. A two-year unpaid leave for the Individual Mobility Grant (European Union, program "Tempus").

Research Grants/Fellowships:

2015 – 2018 - Participation in the grants "Regulated water deficit in forest nurseries" and "Restitution of silver fir in the Sudety Mountains". State Forests National Forest Holding.

2013 – 2016 - Participation in the grant "Secondary sexual dimorphism in dioecious plants. An evolutionary adaptation or a consequence of different strategies of resources allocation? National Science Centre, Poland.

2012 – 2015 Participation in the grant "Adaptive patterns in whole-plant resource allocation, and leaf and root characteristics in temperate woody climbers – a phylogenetically balanced study". National Science Centre, Poland

2010 – 2013 – Director of the grant "Competition between seedlings of Quercus petraea and Prunus serotina growing in different light and allelochemical conditions", Ministry of Sciences and Higher Education.

2009 – 2013 – Participant in the grant "Functional anatomy and ecophysiology of leaves of evergreen woody plants from moderate climate zone". National Science Centre, Poland.

2008 – Scholarship of Kosciuszko Foundation, three month stay at the University of Minnesota, Department of Forest Resources.

2007 – 2001 Participation in "Monitoring of the forest ecosystems in the Karkonosze National Park". The grant has been supported by Polish "Ecofund" and the administration of the Karkonosze National Park (Southern Poland). Ecophysiological research into Abies alba and Fagus sylvatica seedlings used in stands transformation in the lower part of the Sudety Mountains. An evaluation of success performance of silver fir artificial regeneration.

2007 – 1998 "Ecophysiological aspects of stressors effects on forest trees". The study has been supported by the August Cieszkowski Agricultural University of Poznań.

2006 "Long-term changes and ecological fluctuations in the riparian forest". The research project has been supported by the August Cieszkowski Agricultural University of Poznań.

2006 "Acclimation of yew (Taxus baccata L.) needles to irradiance conditions – anatomical plasticity, light absorption and photochemical regulation". The grant from the Adam Mickiewicz University of Poznań and the August Cieszkowski Agricultural University of Poznań. Participation in "Inventory of yew (Taxus baccata L.) in the Forest District "Rokita".

2006 – 1998 "Responses of forest trees at different developmental stages to light stress under different environmental conditions". The study was supported by the August Cieszkowski Agricultural University of Poznań and ended with the post-doctoral (habilitation) thesis.

2002, 1998 two-week stays at the National Institute of Agronomic Research (INRA), Ecophysiology Unit, Nancy, France.

1999, 2000 Post-doctoral Fellowship of the French Government at the National Institute of Agronomic Research (INRA), Ecophysiology Unit, Nancy, France. The research project topic: "Response of silver fir (Abies alba Mill.) seedlings from diverse provenances to several light regimes and drought."

1999 – 2001 Participation in the research program entitled "Reintroduction of silver fir (Abies alba Mill.) into the Karkonosze National Park", the Agricultural University of Poznań. The program was supported by Polish "Ecofund".

1999 Grant from "Polish Committee of Scientific Research" for the purchase of the gas exchange analyser and fluorometer.

1996 Individual Mobility Grant of the European Union (program "Tempus") at the Gembloux Agricultural University, Belgium. A response of Norway spruce (Picea abies Karst.) seedlings from different altitudinal provenances to an enhanced UV-B irradiation was studied. The preparation of the doctoral dissertation.

1993/1994 Individual Mobility Grant of the European Union (the program "Tempus") at the Gembloux Agricultural University, Belgium. The topic of the research project: "Impact of ultraviolet-B radiation on selected provenances of some species of forest trees and vegetal cover".

1992/1993 Individual Mobility Grant of the European Union (the program "Tempus") at the Free University of Brussels for two-semester post-graduate study in molecular biology and biotechnology.

1989 Two-month traineeship in the Second Forest District in canton Neuchâtel, Switzerland. Forest management and silviculture in practice.

Honors:

2019 President of the University Individual Award for publications

2014 President of the University Scientific Team Award for publications

2011 President of the University Scientific Team Award for publications

2009 Team Award of the Ministry of the Environment for the scientific achievements in protection and use of the environment and its resources and for the work "Restitution of silver fir in the Karkonoski National Park".

2006 The letter with honors for habilitation from President of the University, President of the University Award for Habilitation

2004 President of the University Scientific Award "for the original publications"

2004 The letter with honors for supervising of the master thesis nominated for an award from President of the University

1998 President of the University Doctorate Award

Publications from the Last 3 Years:

Trocha L., Weiser E., Robakowski P. 2016. Interactive effects of juvenile defoliation, light conditions, and interspecific competition on growth and ectomycorrhizal colonization of Fagus sylvatica and Pinus sylvestris seedlings. *Mycorrhiza,* 26:47-56. doi 10.1007/s00572-015-0645-4.

Robakowski P., Bielinis E., Mejza I., Bułaj B. 2016. Seasonal Changes Affect Root Prunasin Concentration in Prunus serotina and Override Species Interactions between P. serotina and Quercus petraea. *Journal of Chemical Ecology*, 42(3):202-214. doi 10.1007/s10886-016-0678-y.

Robakowski P., Bielinis E. 2017. Needle age dependence of photosynthesis along a light gradient within an Abies alba crown. *Acta Physiologiae Plantarum,* 39:83. DOI 10.1007/s11738-017-2376-y.

Robakowski P., Bielinis E., Senadall K. 2018. Light energy partitioning, photosynthetic efficiency and biomass allocation in invasive Prunus serotina and native Quercus petraea in relation to light environment, competition and allelopathy. *Journal of Plant Research*, 131:505-523. doi.org/10.1007/s10265-018-1009-x.

Robakowski P., Pers-Kamczyc E., Ratajczak E., Thomas P.A., Ye Z-P, Rabska M., Iszkuło G. 2018. Photochemistry and Antioxidative Capacity of Female and Male Taxus baccata L. Acclimated to Different Nutritional Environments. *Frontiers in Plant Sciences, volume 9,* article 742, doi: 10.3389/fpls.2018.00742.

In: *Prunus*
Editor: William Schneider
ISBN: 978-1-53617-755-8

Chapter 4

A REVIEW OF THE CLASSIFICATION, PROPAGATION TECHNIQUES AND TOXICITY OF ENDANGERED *PRUNUS AFRICANA* (HOOK F.) KALKMAN FOR ITS IMPROVED DOMESTICATION AND CONSERVATION

Justine Germo Nzweundji[1,3], Innocent Afuh Awasom[2], Néhémie Tchinda Donfagsiteli[1,*] and Nicolas Niemenak[3]
[1]Institute of Medical Research and Medicinal Plants Studies, Yaounde, Cameroon
[2]Texas Tech University Libraries, Lubbock, Texas, US
[3]Higher Teacher's Training College, University of Yaounde I, Cameroon

[*] Corresponding Author's Email: donfagsiteli_nehemie@yahoo.com.

ABSTRACT

Prunus africana is among the 400 species from the genus *Prunus* that belong to the subfamily Prunoideae of the family Rosaceae, and the only member of the genus with medicinal value that grows in Africa. Propagation of *Prunus africana* is challenging due to the recalcitrant nature of the seed, whose viability is lost after few months post- harvest; the overexploitation due to the unsustainable harvesting which has threatened the species existence and is therefore listed as an endangered species by the Convention on International Trade in Endangered Species (CITES). Thus, endangered species with medicinal and economical value are listed as priority for domestication in many projects in Africa. *In-situ, ex-situ* and *in-vitro* methods have been set for the conservation which ensure mass productivity and sustainability of the species. Besides the various active components found in the bark of *Prunus Africana* have been used for decades to treat more than 30 human ailments and several diseases in animals. This bark extract is relatively harmless but should be avoid at higher doses. The fresh cut bark, bruised leaves and crushed seeds have a strong cyanide smell and have shown some poisonous effects. This review examines the classification, toxicity, and techniques used to propagate the endangered *Prunus africana* to improve its medicinal utilization and sustainable conservation.

Keywords: phylogenetics, domestication, toxicity, *Prunus africana*. endangered plant species

1. INTRODUCTION

Prunus africana is part of the genus *Prunus* which has more than 250 species distributed across the Northern Hemisphere's tropical and subtropical regions (Lee et Wen 2001). Some of these species are grown for their edible fruit or ornamental plantings, while others for medicinal purposes. *Prunus africana* is the case for *Prunus africana* (Hook. f.) Kalman endemic to Africa and having many local names such as African Cherry, African Prune, Red Stinkwood or *Pygeum* (Hall et al., 2000). It is a large tree of the *Rosaceae* family and belonging to the sub genus Laurocerasus as well as Euro Asiatic and tropical species. The species grows in the Afro-mountainous forests between 1500 and 3000 meters above sea level on volcanic soil and in cool climates. Populations are distributed in the Central, Southern, Eastern and Western regions of Africa, covering 20 countries (Kadu et al., 2011; Cunningham et al., 2014). These populations of *P. africana* are relatively smaller in southern Africa with more representation in West, Central and East Africa, particularly in Cameroon. The bark and leaves traditionally are used to treat many illnesses such as stomach pain, kidney disease, fevers, malaria, arrow poisoning, gonorrhea, and insanity, and also used as appetite stimulant, wound dressing, and purgative (Watt and Breyer-Brandwijk, 1962; Burkill, 1997; Neuwinger, 2000; WHO, 2002; VanWyk et al., 2009; Stewart 2003; Brendler et al., 2010). The tree bark is reputed for its healing properties and treatment against benign prostatic hyperplasia (Ishani et al., 2000).

The major active constituents of *Prunus africana* include docosanol (0.6%) and b-sitosterol (15.7%). Other constituents are alkanols (tetracosanol [0.5%] and *trans*-ferulic acid esters of docosanol and tetracosanol), fatty acids (myristic, palmitic, linoleic, oleic, stearic, arachidic, behenic and lignoceric acids) sterols (sitosterone [2.0%] and daucosterol) and triterpenes (ursolic acid [2.9%], friedelin [1.4%], 2-a-hydroxyursolic acid [0.5%], epimaslinic acid [0.8%] and maslinic acid) (Bruneton 1995; Uberti E et al., 1990).

Many European and U.S. pharmaceutical companies use the tree bark to manufacture more than 19 drugs with potential treatment against benign prostatic hyperplasia (O'Brien, 2000; Cunningham et al., 2002).

Export of *P. africana*'s bark was estimated at more than 4000 tons per year for a value of finished products of about U.S. $220 million (Cunningham et al., 2002). One of the main sources of supply to satisfy this demand are Cameroon's natural populations of *P. africana* with an annual export of bark encrypted around 2000 tons for a total of 1.3 million Euro, approximately (Nsawir and Ingram, 2007). Other participating countries in the international trade are Kenya, Equatorial Guinea and Madagascar. Despite its socio-economic importance, *P. africana* faces many challenges. The international demand of *P. africana* has been unsustainable and exploitative with entire trees being girdled of their bark and left to die or, in other cases, being cut to the ground to facilitate easier access to their bark (Betti 2008). This rapid increase in harvesting the bark of *P. africana* over a relatively short period compelled the Convention on International Trade in Endangered Species (CITES) to classify it as an endangered species in Cameroon (Cunningham et al., 1997) and the International Union for Conservation of Nature (IUCN) to classify it as vulnerable species (IUCN, 2002). Despite the protection afforded by these designations however, *P. africana* remains Cameroon's most intensively exported medicinal plant species by volume (Cunningham et al., 2002; Nsawir et Ingram, 2007; Amougou et al., 2010). A decrease has therefore been noted in its population over the time in terms of tree density, area of occupation, and habitat quality due to the actual level of exploitation (Amougou et al., 2010). The key challenge is therefore to develop suitable methods of propagation that are both appropriate to the species and easily applicable to local communities.

Like most forest tree species, propagation by seeds is the main form of regeneration of *P. africana*. This propagation way is the easiest and cheapest method usually used to regenerate trees (Mapongmetsem, 1994). However, this method of propagation faces two major constraints that hinder the effectiveness of *P. africana's* seeds: the low aptitude of seeds to germinate and the rapid loss of their vigor and viability during storage. Indeed, *P. africana* seeds are dormant due to the inhibition exerted by the pulp and the endocarp. Germination of these seeds removed from their envelopes yielded to date a maximum percentage of germination of 50% (Scandé et al., 2005; Avana, 2006). Unfortunately, these seeds are also recalcitrant, losing their vigor and viability during drying and within less than three months of storage (Sunderland et Nkefor, 1997; Negash, 2002; Avana, 2006). To date, the best germination percentage obtained after six months of storage at 4°C is only of 37% (Avana, 2006).

Another problem is the production of hydrogen cyanide which are bark compounds, very toxic, and sometime creating damages when high doses of crude extracts are taken by local populations. Most African countries where *P. africana* grows have a policy to ensure their sustainable management in the forests. Despite all these management policies in the states, the fact remains that control problems persist. Hence, the development of mechanisms for the identification of raw materials (pharmacognosy and quality control) remains a priority for most producer countries (Kotina et al., 2016). The aim of this review is to examine the classification, chemical composition and toxicity level of crude extract of *Prunus africana* and the techniques used to propagate and disseminate varieties of this endangered plant species so as to improve utilization and sustainable conservation.

2. ORIGIN AND CLASSIFICATION OF *PRUNUS AFRICANA*

2.1. Genus *Prunus*

The genus *Prunus* belongs to the family of Rosaceae and is made up of about 250 species, mainly distributed in the North American, European and Asian continents. Concerning all the species, about 75 come from America and tropical Asia (Lee et Wen 2001). Several species are cultivated for their fruit (apricot, almond, cherry, peach, plum) or for their ornamental value (Japanese and Virginia cherry, laurel-cherry, etc.) (Komarov, 1971; Schery, 1972; Watkins, 1995; Badenes and Parfitt 1995). Some ornamental species with no common name are simply called "*Prunus*." Some species are toxic, for example the cherry laurel. The most important species are: *Prunus armeniaca* L. (Apricot); *Prunus avium* (L.) L. (Cherry, Bigarreautier, Guignier); *Prunus brigantina* Vill. (Apricot or Plum from Briançon); *Prunus cerasifera* Ehrh. (Plum-Cherry, myrobolan); *Prunus cerasus* L. (Griottier); *Prunus domestica* L. subsp. domestica (Plum), subsp italica (Borckh.) Hegi (Reine-claude), subsp. Borck syriaca. (Mirabellier), subsp. Insititia L. (Quetschier); *Prunus dulcis* (Mill.) D. A. Webb (Amandier); *Prunus japonica* Thunb. (Korean cherry); *Prunus laurocerasus* L. (Laurel-cherry); *Prunus lusitanica* L. (Laurel from Portugal); *Prunus maackii* Rupr. (Manchurian Cherry); *Prunus mahaleb* L. (Bois de Sainte Lucie); *Prunus mume* Siebold & Zucc. (Japanese apricot); *Prunus nigra*; *Prunus padus* L. (Birch, stinky wood); *Prunus persica* (L.)

Batsch (Peach); *Prunus prostrata* Labill. (Prostrate cherry); *Prunus salicina* Lindl. (Japanese plum); *Prunus serotina* Ehrh.; *Prunus serrula* Lindl. (Tibetan cherry); *Prunus serrulata* Lindl. (Japan cherry tree); *Prunus spinosa* L. (Blackthorn); *Prunus tenella* Batsch (Dwarf Almond); *Prunus virginiana* L. (Virginia Cherry); Prunus × cerea (L.) Ehrh. (*P. cerasifera* × *P. domestica*) (Mirabellier); Prunus × fruticans Weihe (Large blackthorn). The genus *Prunus* is divided into 2 classes (Amygdalus-Prunus, Cerasus-Laurocerasus-Padus); 4 subgenera [Amygdalus (almond and peach trees), *Prunus*, Cerasus and Emplectocladus] and 7 sections (Armeniaca (apricot trees), Penarmeniaca, Prunocerasus, *Prunus* (plum trees), Cerasus (cherry trees), Laurocerasus (cherry laurels)), Microcerasus)). This classification was confirmed by the work of Wen et al., (2008) using the sequences of the ITS regions of the ribosomal nucleus and of the chloroplast gene and thus demonstrating the dynamism and the complexity of this genus. *Prunus africana* is one of the most important species of the genus *Prunus* because of its use in traditional medicine and the use of these barks for the manufacture of medicines, in particular against treatment of benign prostatic hyperplasia (Cunningham et al., 2002).

2.2. Origin and Geographical Distribution of *Prunus africana*

Prunus africana is native to Africa as indicated by its specific epithet "africana" (Graham, 1960). It is a dominant endemic species of the Afro-mountain forests of Africa and Madagascar. In addition, it grows between 900 and 3000 meters above sea level, but more precisely between 1000 and 2500 meters. However, studies have shown that it grows as well in transitional forests between lowlands and mountain regions (Cunningham and Mbenkum, 1993). Other studies indicate its presence at altitudes as low as 600 meters at least (Geldenhuys, 1981; Mbile et al., 2003). The distribution of *P. africana* is related to its preference for annual temperatures between 18° and 26°C and with average rainfall of 2000 mm (Achoundong, 1995). Hall et al., (2000) posit that this species is present in about twenty African countries from Ethiopia to South Africa, and from Nigeria to Madagascar. It is also present in some twenty forests in countries of Eastern, Central and Southern Africa.

Table 1. Distribution of *Prunus africana* in Range State (Bruneton 1995)

Countries	Localities
Angola	Bailundu highlands region and Mt. Moco
Burundi	Montane forest (e.g., Teza forest), Albertine Rift, Probably found in Mt. Heha/Ijenda, Mt. Bururi
Cameroon	North West region highlands (Mt Kilum, Oku, Mt. Manenguba, Adamawa plateau and South Est region (Mt. Cameroon)
DR Congo*	Mountains region of Rwenzori and Virunga, Kivu region, National Park of Kahuzi-Biegaand probably on Itombwe
Equatorial Guinea	Island of Bioko (Pico Basilé and Grand Caldera de Luba)
Ethiopia	Regions from 1500-2300 m. North West to South East highlands from lake Tana to Harar; Montane and valley forests of Harerge, Illubabor, Kefa, Arsi, Wolega
Kenya	High lands (Mt Elgon and Mt. Kenya) and Mau forests
Lesotho	One collection from Rock pools area, Sehlabathebe. One specimen reported from Maphotong Gorge (2)
Madagascar	Moist Montane forests (1000-2000m asl) (eg. Zahamena Strict) Nature Reserve, Mantadia, Antsevabe and Manakambahiny-Est
Malawi	Mt Mulanje, Zomba and Vipya planteaus
Mozambique	High lands (Mt Gorongosa and Chiperone), Chimanimani mountains.
Nigeria	Mambila plateau, South East of Nigeria
Rwanda	High lands (Virunga mountains) and forest (Mukura and Nyungwe)
Sao Tome e Principe	Central Principe (volcanic plugs of Joao Dias Pai e Filho and montane Sao Tome from 1200-1400m asl)
South Africa	Afromontane forest patches from Mpumalanga through KwaZulu/Natal to the Knysna forest
Sudan	Imatong mountains
Swaziland	Forest patches near Malolotja (Forbes Reef) and Mbabane.
Tanzania	Moist evergreen forests in North East Tanzania, including Mt Kilimanjaro
Uganda	South West high lands Uganda (Kalinzu, Bwindi, Mgahinga and Mt. Elgon, Imatong mountains)
Zambia	Relict forest patches in fire maintained upland grasslands
Zimbabwe	Chimanimani mountains and Inyanga

It is in eastern countries such as Ethiopia, Kenya, Rwanda, Sao Tome and Principe and Madagascar. And found in the central countries of Cameroon, Democratic Republic of Congo, Angola, Burundi, Equatorial Guinea. In Southern Africa it is found in South Africa, Sudan, Swaziland, United Republic of Tanzania, Lesotho and Mozambique. (CITES, 2007) (Table 1).

Cameroon has a large *P. africana* population spread across more than eighty sites, thus covering six of the ten regions of the country (Mbile et al., 2003). This species is distributed in three major regions distinguished by volcanic soils and a cold climate of the regions of high altitude (Cunningham and Mbenkum, 1993; Simons et al., 1998). The *Prunus africana* sites in Cameroon are mainly Mount Cameroon, Mount Oku, and the highlands of West Cameroon.

2.3. Classification

Prunus africana, plant of the Rosaceae family is described for the first time by Hooker (1864) under the name of *Pygeum africanum* then attached to the genus *Prunus* by Kalkman who considered 22 groups of the genus of the Rosaceae family and subdivided them into four subfamilies as seen in Figure 2. Thorne (1992) in his classification, considered five sub-families, namely Spiraeoideae, Rosoideae, Maloideae and Amygdaloideae which replaced Prunoideae in addition to Quillajoideae. On the basis of phylogenetic analysis from variations in the sequences of chloroplast genes, the integrity of the three subfamilies Rosoideae, Maloideae and Amygdaloideae has been confirmed (Morgan et al., 1994). These authors mark a clear distinction between the Maloideae and Amygdaloideae and the other groups suggesting that the latter were older. The species of *Prunus* have two subgenera: *Padus* containing the deciduous species and *Laurocerasus* the evergreen species (Kalkman, 1965). *Prunus africana* in the same way as the Asian and tropical Asian species of the same genus has been included in the subgenus of *Laurocerasus* (Figure 1).

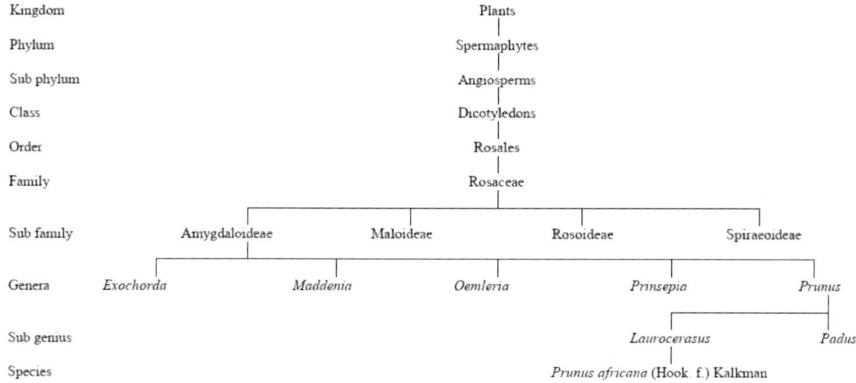

Figure 1. Classification of *Prunus africana* within the family Rosaceae.

2.4. Populations of *Prunus africana*

Several studies have been conducted on the genetic diversity of *P. africana* populations in Africa using various molecular markers RAPD, AFLP and microsatellites (Dawson and Powell 1999; Avana et al., 2004; Cavers et al., 2009). Among the determining factors for a conservation strategy, the evaluation of the genetic diversity of the different populations of the species is an important factor. Whether in Cameroon, Ethiopia, Kenya or other countries, these markers have shown a high degree of diversity between populations. In Cameroon, two groups have been distinguished, corresponding to two of the massifs that make up the Cameroonian ridge (Avana et al., 2004°). In Ethiopia, studies on 21 populations using plastid microsatellites have helped to better understand the phylogeography of the species in order to develop the best conservation strategies. This is how 16 plastid haplotypes have been detected in the population and having a higher degree of diversity than that of accessions from other African origins (Mihretie et al., 2015). Ten of these 16 plastid holotypes were unique suggesting an isolated evolution of the Ethiopian plastids in opposition to other origins whose evolution is divergent (Mihretie et al., 2015).

3. IMPORTANCE OF *PRUNUS AFRICANA*

Many studies have reported the uses of *Prunus africana* as a traditional medicine (Burkill, 1997). *Prunus africana* (Hook. f) Kalkman is a multipurpose tree, that has been used locally, nationally, regionally and internationally mostly for its pharmaceutical properties (Schippmann, 2001). The species is classified among the very important tree species. All parts of the tree have medicinal value, but the main use is the bark extract for the treatment of prostate gland hypertrophy (PGH) and benign prostatic hyperplasia (BPH) that affects 50% of men above age 50. In many countries in Africa, traditional healers use *P. africana* for the treatments of more than 30 ailments and animal diseases such as fevers, epilepsy malaria, arthritis infertility, stomach ache, diarrhea, eye disorders, impotence, syphilis kidney disease, obesity, pneumonia, hypertension, asthma, urinary problems, mental disorders and diseases related to prostate cancer (Stewart, 2003, Amougou et al., 2010). It is equally used in wound dressings (orexigenic properties) to improve the appetite as well as purgative for cattle. The tree is also used as source of timber for furniture, for charcoal, window and doorframes, wagon fuel, furniture, bridges and house building and in many countries the timber is used for heavy construction works (Stewart, 2003). *Prunus africana* is a high economic species in Cameroon especially in the rural communities where it grows. People living around Mount Cameroon get about 70% of their annual cash income from this activity, and 60% of households in this area are involved in *P. africana* harvesting and trade. It thus represents a major secondary source of income. Its exploitation started in 1970s in Cameroon (Ingram et al., 2009). Since 1995, the plant has been listed by the Convention on International Trade in Endangered Species (CITES) as an endangered species. The species usually grows in high altitude and is the most popular medicinal species that is used by local population in the mountain forest areas. Its exportation and consequent economic importance has been well documented, and its benefits impact most from the local population to industries. In the past few decades the industry has witnessed the production expand from local to a large scale that involves national as well as international trade (Vinceti et al., 2013). About 4000 tons per year of *Prunus africana* bark is exported with an estimated value of about $ 220 million (Cunningham et al., 2002). In general, the exploitation and domestication of the species are used to sustain the economy and environment of some developing countries, while reducing the threat and the extinction of wild population (Giuliani et al., 2005).

4. Pharmaceutical Impacts

The pharmaceutical impacts of *Prunus africana* has increased over the past decades and have been used for the preparation of over 40 brand-name products. The most commonly known use is to treat prostate disorders, in particular Benign Prostatic hyperplasia which affects 50% of men above the age of 50. The leaves are used to treat fever while the infusion improves appetite. The pounded bark is used to treat stomachache and also as purgative for cattle. The different active components of the species essentially play a fundamental role in the plant's defense to fight against herbivory. The bioactivities in the plants are generally ascribed to the presence of plant secondary metabolites which could have beneficial or adverse effects (Makkar et al., 2007).

5. Mechanisms of Action of Active Components

The different active components found in *Prunus africana* are listed in six different major groups: n-docosanol, beta-sitosterol, alkanols, fatty acids, sterols and triterpenes (Catalano et al., 1984, Bruneton 1995, Bombardelli and Morazzoni, 1997, Kadu et al., 2012). The active compounds have not yet been synthesized. Phytosterol found in all plant food is related structurally to cholesterol and reduce the level of its serum (Rocha et al., 2011). Studies have shown that phytosterols (e.g., beta-sitosterol) have anti-inflammatory effects that block the production of pro-inflammatory prostaglandins in the prostate. In Cameroon, there is variation of β-sitosterol from zero to 38.65 μg/ml with respect to altitude and chemical elements of the soils (Tchouakionie et al., 2014). The pentacyclic triterpenes (ursolic and oleanic acids) have many antitumor, anti-eadema and anti-inflammatory effects (Kvasnica et al., 2015). Another major component are ferulic acid nesters (n-docosanol and tetracosanol), derivatives of ferulic acids with the role in the reduction of prolactin quantity that inhibited the accumulation of cholesterol in the prostate. Prolactin is purported to increase the uptake of testosterone by the prostate, and cholesterol increases binding sites for dihydrotestosterone (Simons et al., 1998; Murray, 1995). *Prunus africana* through the synergistic action of bioactive components has anti-inflammatory activity, by decreasing production of leukotrienes and other 5-lipoxygenase metabolites. The species also has the ability to kill cell tumors and prevent the proliferation of prostate cancer (Komakech et al., 2017). Patients of BPH receiving treatment from bark extract experience a considerable reduction in frequency of urination and residual urine volume compared to the placebo (Quiles et al., 2010). The synergistic effects of phytosterols (anti-inflammatory, anti-estrogen), pentacyclic triterpenes (anti-edema, immunostimulant) and ferulic acid esters (hypocholesterolaemic) lead in general to the reduction in oedema, inflammation and histamine-induced vessel permeability (Quiles et al., 2010).

6. Toxicity Effects

The wood of *P. africana* as well as the bruised leaves and fruits have a strong bitter-almond smell mostly when first cut and bitter taste; this indicates the presence of hydrogen cyanide (WHO, 2002). The hydrogen cyanide is colorless and extremely poisonous. Different degrees of toxicity have been observed on *P. africana* extracts. *In vitro* and *in vivo* studies have shown strong pharmacological effect to kill tumor cells via apoptotic pathways, preventing the proliferation of prostate cancer cells. Thus, the effect of *P. africana* active components could be an effective cytotoxic chemotherapy for the treatment of diseases related to prostate cancer. Different extracts from *P. africana* bark showed different level of toxicity. The bark extract from chloroform on mice and rats shows that there is no side effects (ESCOP 2009; Bombardelli and Morazzoni 1997) when administrated short-term. However, long-term administration caused clinical signs of toxicity, damage of organs and a 50% mortality rate. The bark extract from dichloromethane and methanol in the *Salmonella typhimurium* TA98 strain showed no mutagenic effect and high toxicity when applied many times (Elgorashi et al., 2003). Mutagenic and clastogenic effect of *in vivo* and *in vitro* studies were not observed (ESCOP 2009). The methanolic bark extract of *P. africana* on rats is not harmful and no mortality observed even with high dose of administration, including no observed carcinogenicity.

7. Domestication and Propagation Technique

Multiple regional and national research programs have classified *Prunus africana* as among the top medicinal plants to be domesticated in Africa. Two of note are the domestication program of agroforestry species conducted by International Center for Research in Agroforestry (ICRAF), Kenya (Tchoundjeu et al., 2002b) and domestication program of medicinal plants conducted by the Institute of Medical Research and Medicinal Plants Studies Cameroon (IMPM 2019). Priority setting of medicinal plants is based on their socio-economic importance with ethnobotanical uses, commercial value, biological criteria and their conservation status. The domestication involves the selection of accessions and development of propagation techniques. Currently, propagation by seeds is still the only technique used to produce seedlings for regeneration programs. The seeds of species are hardy and hard-to-germinate and thus their viability is lost after few months post-harvests, but little is known about mechanism occurring during germination. However, the characterization aspects of *P. africana* seed during germination using morphological, histological and biochemical tools was set to develop a reliable protocol for regeneration of the species after long term conservation (Scandé et al., 2005; Nzweundji et al., 2019). In addition, Nzweundji et al., (2019) showed that seeds of *P. africana* are polyembryonic. This characteristic is useful to improve seed multiplication and enables biotechnologists to get cultivars/progenies having parents' characteristics. Multiplication by cutting has been initiated by Tchoundjeu et al., (2002a). If the percentage of rooting (80%) was high, the success of whole plant regeneration and transplantation in the farm or agroforestry system was still very low. In addition, the domestication of species is challenging during in situ regeneration process and this is sometimes related to the lack of mycorrhizal fungi in the soil of the original plant. These soil microorganisms operate in a symbiotic relationship by facilitating the absorption of mineral nutrients to the plants while receiving organic carbon in the form of glucose and lipid (Smith and Read, 2008; Keymer and Gutjahr; 2018). Isolation and use of those indigenous fungi are necessary to the success of the large-scale cultivation and its introduction in the agroforestry system. Studies have shown some specificities of fungi associated with *P. africana* in Cameroon (Wubet et al., 2003; Tchiechoua, 2012; Nzweundji et al., 2017). Mycorrhizal fungi will be required for a successful enrichment of the plantations of *P. africana* species; they are biofertilizer, they improve the quality, the productivity of the species as well as improvement of yielding and active components (Kadu et al., 2012 Tchiechoua et al., 2019). As the seeds of *Prunus africana* are not available all year long due to sporadic fructification and flowering (Dawson et al., 2000), the use of another tree parts for propagation seems necessary. Tissue culture via micro cutting is promising and is another useful alternative technique for more rapid clonal multiplication of medicinal species (Donfagsiteli et al., 2013a,b; Nzweundji et al., 2016). After one month of culture in Murashige and Skoog solution supplemented with growth regulators, a piece of seeds produced more than 50 somatic embryos which could produce about 50 young plants (Nzweundji, 2016). Tissue culture techniques help to improve *P. africana* micropropagation with the potential of enhancing propagation rate. This opens a new window of opportunities for researchers, medical firms and farmers and all those involved in *Prunus africana* production for genetic improvement research and large-scale cultivation.

Conclusion

Prunus africana an endemic species, is a typical African plant that grows in several countries, mainly in mountainous areas. Genetic variability, according to provenances, has been established. Due to its socio-economic importance, the species is one of the most important medicinal plants in most domestication and conservation programs in Africa. With increasing overexploitation, the first conservation measures consisted in developing the propagation methods. Thus, seed multiplication is the technique currently used for seed production although vegetative propagation tests have been undertaken. The presence of certain toxic compounds in plant extracts is not without risk for local African populations who use bark decoctions, without dose control measures, to traditionally treat prostate hyperplasia and other ailments. Although regulations exist in African countries on the exploitation and marketing of the species, we recommend that measures be taken to the medicinal composition and dosage in the various pharmacopoeias in order to strengthen and guarantee its use for therapeutic purposes.

References

Achoundong, G. 1995. *Prunus africana*. Rosaceae, Forest tree to discover. *Woods and Forests of tropics* 245p.

Amougou, A., Betti J. L., Ewusi N. B., Mbarga, N., Akagou, Z. H. C., Fonkoua, C., Essomba, E. R. and Nkouna A. C. (2010). Preliminary report on sustainable harvesting of *Prunus africana* (*rosaceae*) in the North West region of Cameroon. *Report prepared for the National Forestry Development Agency (ANAFOR), the Cameroon CITES*

Scientific Authority for flora, in the frame of the project "Non-detriment findings for Prunus africana (Hook. f.) Kalman in Cameroon."

Avana, M. L. (2006). Domestication of *Prunus africana* (Hook. F.) *kalkmam (Rosaceae): Study of germination and cutting*. PhD thesis, University of Yaounde I. 132p.

Avana, M. L., Tchoundjeu, Z., Bell, J. M., Alexandre, V. and Chevallier, M. H. (2004). Genetic diversity of *Prunus africana* (Hook. f.) Kalkman in Cameroon. *Woods and Forests of tropics*, 4: 41-49.

Badenes, M. L. and Parfitt, D. E. (1995). Phylogenetic relationships of cultivated *Prunus* species from an analysis of chloroplast DNA variation. *Theoretical and Applied Genetics*, 90: 1035-1041.

Betti, J. L. (2008). Non-Detriment Findings Report on *Prunus africana* (Rosaceae) in Cameroon. *Report prepared for the International Expert Workshop on Non-Detriment Findings*, Mexico, 52 p.

Bombardelli, E. and Morazzoni, P. (1997). *Prunus africana* (Hook. f). *Fitoterapia*, 68:205-218.

Brendler, T., Eloff, J. N., Gurib-Fakim, A., Phillips, L. D. (2010). *African Herbal Pharmacopoeia. Association for African Medicinal Plant Standards and Graphic Press*, Mauritius.

Bruneton, J. (1995). *Pharmacognosy, Phytochemistry, medicinal plants*. Lavisior Paris.

Burkill, H. M. (1997). *The Useful Plants of West Tropical Africa 4* (2nd ed.), Families M-R. Royal Botanic Gardens, Kew.

Catalano, S., Ferretti, M., Marsili, A. and Morelli, I. (1984). New constituents of *Prunus africana* bark extract. *Journal of Natural Products*, 47: 910.

Cavers, S., Munro, R. C., Kadu, C. A. C. and Konrad, H. (2009). Transfer of microsatellite loci for the tropical tree *Prunus africana* (Hook. f.) Kalkman. *Silvae Genetic*, 58 :276-279.

CITES. (2007). *Conservation newsletter on international trade in endangered species of wild flora and fauna in Africa, Volume 1*, N° 1. SSN, 6p.

Cunningham, A. B., Ayuk, E., Franzel, S., Duguma, B., Asanga, C. (2002). An Economic Evaluation of Medicinal Tree Cultivation: Prunus africana in Cameroon. *People and Plants Working Paper 10.* UNESCO, Paris.

Cunningham, A. B., Campbell, B. M., Luckert, M. K. (2014). *Bark use, Management, and trading in Africa. Advances in Economic Botany 17.* The New York Botanical Garden Press, New York.

Cunningham, A. B. and Mbenkum F. T. (1993). Sustainability of harvesting *Prunus africana* bark in Cameroon. A medicinal plant in international trade. UNESCO, Paris. *People and plants working paper 2*. 28 p.

Cunningham, M., Cunningham, A. B. and Schippmann, U. (1997). *Trade in Prunus africana and the Implementation of CITES*, German Federal Agency for Nature Conservation, Bonn, Germany.

Dawson, I. K. and Powell, W. (1999). Genetic variation in the Afromontane tree *Prunus africana*, an endangered medicinal species. *Molecular Ecology*, 8: 151-156.

Donfagsiteli, T. N., Mbita, M. H. J. C., Fotso, Nzweundji, J. G., Oumar, D., Dongmo, B. Sanonne, Agbor, A. G. and Omokolo N. D. (2013a). Biochemical aspects of single-node cuttings of *Ricinodendron heudelotii* (Baill.) in relation with rooting. *African Journal of Biotechnology,* 12(X):1049-1056.

Donfagsiteli, T. N., Mbita, M. H. J. C., Fotso, Nzweundji, J. G., Tsabang, N., Dongmo, B., Oumar, D., Tarkang, P. A., Carver, A. and Omokolo N. D. (2013b). Improving propagation methods of *Ricinodendron heudelotti* (Baill) from cuttings. *South African Journal of Botany*, 88: 3-9.

Elgorashi, E. E. 1., Taylor, J. L., Maes, A., van Staden, J., De Kimpe, N. and Verschaeve, L. (2003). Screening of medicinal plants used in South African traditional medicine for genotoxic effects. *Toxicology Letter*, 43(2):195-207.

ESCOP. (2009). *Monograph*, European Scientific Cooperative on Phytotherapy (ESCOP) Notaries House, Chapel Street, Exeter EX1 1EZ, United Kingdom.

Geldenhuys, C. J. (1981). *Prunus africana* in the Bloukrans River Gorge, Southern Cape. *South Africa Journal of Forestry*, 118: 61- 66.

Giuliani, A., Nashwa, A. and Markus, B. (2005). *Linking Biodiversity Products to Markets to improve the Livelihoods of the Resource Poor: Case Study on the Market Chain of Capers in Syria*, Rome, Italy.

Graham, R. A. (1960). Flora of Tropical East Africa. *Crown Agents for Overseas, Government and Administration.* London. 61p.

Hall, J. B., Eileen, M., O'Brien, Fergus, L. and Sinclair. (2000). *Prunus Africana A Monograph*, School of Agricultural and Forest Sciences, University of Wales, Bangor, U.K.

Hall, J. B., O'Brien, E. M. and Munjuga, M. (2000). Ecology and Biology. In: Hall, J. B., O'Brien, E. M., Sinclair, F., (eds). *Prunus africana*: *A Monograph School of Agricultural and forest Sciences,* University of Wales, Bangor. pp. 53-59.

IMPM 2019 New approach to prioritization of medicinal plants for domestication in the development program of medicinal products. *Report Project Domestication of medicinal plant*, Cameroon 56p.

Ingram, V., Owono, A., Schure, J. and Ndam, N. (2009). *Guidance for a national Prunus africana management plan*, Cameroon. CIFOR, FAO.

Ishani, A., MacDonald, R., Nelson, D., Rutks, I., Wilt, T. J. (2000). *Pygeum africanum* for the treatment of patients with benign prostatic hyperplasia: a systematic review and quantitative meta-analysis. *The American Journal of Medicine*, 109:654–664.

IUCN, *2002 IUCN Red List of Threatened Species*. Available from <http://www.redlist.org>. Downloaded on 7 June 2011.

Kadu, C. A. C., Parich, A., Schueler, S., Konrad, H., Muluvi, G. M., Eyog-Matig, O., Muchugi, A., Williams, V. L., Ramamonjisoa, L., Kapinga, C., Foahom, B., Katsvanga, C., Hafashimana, D., Obama, C., Vinceti, B., Schumacher, R. and Geburek, T. (2012). Bioactive constituents in *Prunus africana*: Geographical variation throughout Africa and associations with environmental and genetic parameters. *Phytochemistry*, 83: 70-78.

Kadu, C. A. C., Schueler, S., Konrad, H., Muluvi, G. M. M., Eyog-Matig, O., Muchugi, A., Williams, V. L., Ramamonjisoa, I., Kapinga, C.,

Foahom, B., Katsvanga, C., Hafashimana, D., Obama, C. and Geburek, T. (2011). Phylogeography of the Afromontane *Prunus africana* reveals a former migration corridor between East and West African highlands. *Molecular Ecology*, 20:165-178.

Kalkman, C. (1965). The Old-World species of *Prunus* sub-genus *Laurocerasus* including those formerly referred to as *Pygeum*. *Blumea*, 13: 1-115.

Keymer, A. and Gutjahr, C. (2018). Cross-kingdom lipid transfer in arbuscular mycorrhiza symbiosis and beyond. *Current Opinion Plant Biology*, 44: 137-144.

Komakech, R., Kang, Y., Lee, J. H. and Omujal, F. (2017). A review of the potential of phytochemicals from *Prunus africana* (Hook f.) kalkman stem bark for chemoprevention and chemotherapy of prostate cancer. *Evidence Based Complement. Alternative Medicine*, doi:10.1155/ 2017/3014019.

Kotina, E.L., Oskolski, A.A., Tilney, P.M. and VanWyk B.E. (2016). Bark and wood structure of *Prunus africana* (Rosaceae), an important African medicinal plant. *South African Journal of Botany*, http://dx.doi.org/10.1016/j.sajb.2016.04.015.

Komarov, V. L. (1971). Rosaceae-Rosoideae, Prunoideae. In: *Flora of the U.S.S.R. Vol. 10* (English translation). Washington, D.C.: Smithsonian Institution.

Kvasnica, M., Takács, B., Holaza, J. and kIngole, D. (2015). Reachability Analysis and Control Synthesis for Uncertain Linear Systems in MPT. *IFAC-Papers OnLine* 48 (14):302-307.

Lee, S. and Wen, J. (2001). A phylogenetic analysis of *Prunus* and the Amygdaloideae (Rosaceae) using ITS sequences of nuclear ribosomal DNA. *American Journal of Botany*, 88: 150–160.

Makkar, H. P. 1., Siddhuraju, P. and Becker; K. (2007). Plant secondary metabolites. *Methods Molecular Biology*, 393:1-122.

Mapongmetsem, P. M. (1994). *Phénologie et modes de propagation de quelques essences locales à potentiel agroforestier en zone forestière*. Thèse de troisième cycle. Université de Yaoundé 176p. [*Phenology and methods of propagation of some local species with agroforestry potential in forest areas*. Postgraduate thesis. University of Yaoundé 176P].

Mbile, P., Tchoundjeu, Z., Degrande, A., Assah, E. and Nkuikeu R. (2003). Mapping the biodiversity of the "Cinderella trees" in *Cameroon. Biodiversity*, 4(2):17-21.

Morgan, D. R., Soltis, D. E. and Robertson, K. R. (1994). Systematic and evolutionary implications for rbcL sequence variation in Rosaceae. *American Journal of Botany*, 81: 890-903.

Mihretie, Z., Schueler, Konrad, H. Bekele, E. and Geburek, T. (2015). Patterns of genetic diversity of *Prunus africana* in Ethiopia: hot spot but not point of origin for range-wide diversity. *Tree Genetics and Genomes*, doi 10.1007/s11295-015-0945-z.

Murray, M. T., 1995. *The Healing Power of Herbs*, 1. Prima Publishing 286–293.

Negash, L. (2002). Review of research advances in some selected African trees with special reference to Ethiopia. *Ethiopian Journal of Biological Science*, 1: 81-126.

Neuwinger, H. D. (2000). *African Traditional Medicine: A Dictionary of Plant Use and Applications*. Medpharm Scientific, Stuttgart.

Nsawir, A. T. and Ingram, V. (2007). *Prunus africana*: Money growing on trees? A plant that can boost Rural economies in Cameroon highland. *FAO Nature & Faunal Journal issue 22*. The value of Biodiversity.

Nzweundji, J. G., Konan, K., Nyochembeng, L. M., Donfagsiteli, T. D. and Niemenak, N. (2019). Germination improvement in threatened medicinal *Prunus africana* for its better domestication: effect of temperature, growth regulators and salts, *Journal of forestry research*, doi 10.1007/s11676-019-01000-0.

Nzweundji, J.G. (2016). *Tissus culture of Prunus africana for its better conservation and domestication in Cameroon agroforestry system*. PhD Thesis, University of Yaounde I; 150p.

O'Brien, E. M. (2000). Pharmaceutical products. In: Hall J. B., O'Brien E. M., Sinclair F., (eds). *Prunus africana: A Monograph*. School of Agricultural and forest Sciences, University of Wales, Bangor, U.K., 53-59.

Quiles, M. T., Arbós, A. A., Fraga, A., de Torres, I. M., Reventós, J., Morote, J. (2010). Antiproliferative and apoptotic effects of herbal agent *Pygeum africanum* on cultured prostate stromal cells from patients with benign prostatic hyperplasia (BHP). *Prostate*, 70: 1044-1053.

Rocha, M. 1., Banuls, C., Bellod, L., Jover, A., Victor, V. M. and Hernandez-Mijares, A. (2011). A review on the role of phytosterols: new insights into cardiovascular risk. *Current Pharmaceutical Design*, 17:4061- 4075.

Scandé, M., Pritchard, H. W. and Dudley, A. E. (2005). Germination and storage characteristics of *Prunus africana* seeds. *News forests*, 27:239-250.

Schery, R. W. (1972). *Plants for man*. 2nd ed. New Jersey: Prentice Hall.

Schippmann, U. (2001). Medicinal plants significant trade. *CITES Project S - 109, Plants.*

Simons, A. J., Dawson, I. K., Dugumba, B. and Tchoundjeu, Z. (1998). Passing problems: prostate and *Prunus*. *Herbal* Gram, 43:49-53.

Smith, S. E. and Read, D. (2008). *Mycorrhizal Symbiosis*, 3rd ed.; Elsevier: Amsterdam, Netherlands, ISBN 9780123705266. 20.

Stewart, K. M. (2003). The African cherry (*Prunus africana*): can lessons be learned from an over-exploited medicinal tree? *Review Journal of Ethnopharmacology*, 89(I):3-13.

Sunderland, T. and Nkefor, J. (1997). Conservation through cultivation. A case study: the propagation of *Prunus africana*. *TAA* Newsletter, pp. 5-12.

Tchiechoua, Y. H. (2012). Morphological and molecular studies of arbuscular mycorhizal fungi associated to *Prunus africana* (Hook. f.) *Kalkman in two agroecological region in Cameroon*, Master Thesis, University of Yaounde, Cameroon.

Tchiechoua, Y. H., Kinyua, J., Wambui, N. V., and Warambo, O. D. (2019). Effect of Indigenous and Introduced Arbuscular Mycorrhizal Fungi on Growth and Phytochemical Content of Vegetatively Propagated *Prunus africana* (Hook. f.) Kalkman Provenances. *Plants*, 2020: 9-37.

Tchouakionie, M., Youmbi, E., Ndoumbe, N. M., Kouam, S. F., Lamshôft, M. and Spiteller, M. (2014). Effect of some abiotic factors on the concentration of β- sitosterol of *Prunus africana* (Hook. f.) Kalkman in the tropical forests of Cameroon. *International Journal of Agronomy and Agricultural Research*, 4 (2):1-11.

Tchoundjeu Z., Avana M. L., Leakey R. R. B., Simons A., Asaah E., Duguma B. and Bell J. M. (2002a). Vegetative propagation of *Prunus africana*: effects of rooting medium, auxin concentration and leaf areas. *Agroforestry Systems*, 54:183-192.

Tchoundjeu Z., Tonye J. and Anegbeh P. (2002b). Domestication of key indigenous non-timber forest products: their economic and environmental potentials in degraded zone of West and Central Africa, In: Kengue J., Kapseu C. et Kayem G. J. (Eds.). *Valorisation du safoutier et autres oléagineux non conventionnels*. Yaoundé (Cameroun). pp. 51-59.

Thorne, R. F. (1992). Classification and geography of the flowering plants. *Botanical Review*, 58: 225-348.

VanWyk, B. E., Van Oudtshoorn, B. and Gericke, N. (2009). *Medicinal Plants of South Africa*. Second ed. Briza Publications, Pretoria.

Vinceti, B., Loo, J., Gaisberger, H., van Zonneveld, M. J., Schueler, S., Konrad, H., et al., (2013). Conservation Priorities for *Prunus africana* Defined with the Aid of Spatial Analysis of Genetic Data and Climatic Variables. *Plos one*, 8(III): e59987.

Watt, J. M. and Breyer-Brandwijk, M. G. (1962). *The Medicinal and Poisonous Plants of Southern and Eastern Africa*. second ed. Livingstone, London.

Watkins, R. (1995). Cherry, plum, peach, apricot and almond. In: Smartt J, Simmonds NW eds. *Evolution of crop plants*. 2nd ed. Burnt Mill: Longman Scientific and Technical.

Wen J. Bergren, ST. Lee CH. Ikerd-Bond S. Yi TS. Yoo KO. Xie L. Shaw J and Pottre D (2008). Phylogenetic inference of Prunus (Rosaceae) using chloroplastic ndhF and nuclear ribosomial ITS sequences. *Journal of Systematics and Evolution* 46: 322-332.

WHO. (2002). *WHO Monographs on Selected Medicinal Plants 2*. World Health Organization, Geneva.

Wubet, T., Kottke, I. Teketay, D. and Oberwinlcer, F. (2003). Mycorrhizal status of indigenous trees in dry Afromontane forests of Ethiopia. *Forest Ecology and Management*, 179: 387-399.

INDEX

A

B

C

D

E

F

G

H

N

O

P

Q

R

S

T

U

V

W